COLLISIONS

ALSO BY ALEC NEVALA-LEE

NONFICTION

Astounding:
John W. Campbell, Isaac Asimov, Robert A. Heinlein, L. Ron Hubbard, and the Golden Age of Science Fiction

Inventor of the Future:
The Visionary Life of Buckminster Fuller

NOVELS

The Icon Thief

City of Exiles

Eternal Empire

SHORT FICTION

Syndromes: Science Fiction Stories

COLLISIONS

A Physicist's Journey from Hiroshima to the Death of the Dinosaurs

Alec Nevala-Lee

W. W. NORTON & COMPANY

Independent Publishers Since 1923

Image on previous spread:
Ten-inch bubble chamber photo of muon-catalyzed fusion.
The American Physical Society.

Printed in the United States of America

First Edition

For information about permission to reproduce selections from this book, write to
Permissions, W. W. Norton & Company, Inc., 500 Fifth Avenue, New York, NY 10110

For information about special discounts for bulk purchases, please contact
W. W. Norton Special Sales at specialsales@wwnorton.com or 800-233-4830

Manufactured by Lakeside Book Company
Design by: Brian Mulligan
Production manager: Louise Mattarelliano

ISBN 978-1-324-07510-3

W. W. Norton & Company, Inc., 500 Fifth Avenue, New York, NY 10110
www.wwnorton.com

W. W. Norton & Company Ltd., 15 Carlisle Street, London W1D 3BS

10 9 8 7 6 5 4 3 2 1

To Wailin

Gossip is the backbone of physics.

—Attributed by Luis W. Alvarez
to J. Robert Oppenheimer

CONTENTS

COLLISIONS

PROLOGUE

Successful physics is basically antisocial, in that it involves showing where the most illustrious physicists of the past have been wrong.

—Luis W. Alvarez

On May 31, 1970, the physicist Luis W. Alvarez took his family on a field trip to the San Leandro Rifle Range. He parked his car just outside a fenced wedge of overgrown grass and scrub that covered half an acre of shoreline, where a narrow sliver of San Francisco Bay extended into the land like a pointing finger. It was a pleasant drive of about half an hour from his workplace at the Lawrence Radiation Laboratory in Berkeley, which made it a convenient location for him to dramatically demonstrate his theory about the death of a president.

As Alvarez emerged from the driver's seat, the weather seemed perfect for a Sunday excursion. He was fifty-eight and casually dressed in a collared shirt with short sleeves, which would have made him look equally at home in an accountant's office or at a console for the launch of a spacecraft. A lean six foot two, he was no longer the skilled gymnast of his youth—most of his exercise these days was confined to the golf course—but he was still an imposing figure, with fine blond hair receding from a high forehead and piercing blue eyes behind browline glasses.

Beside him was Janet, his second wife, who was two decades younger and smartly attired in shades, a plaid sunhat, and a sleeveless blouse. In her arms Janet carried Helen, their two-year-old daughter, while their son, Donald, who was four, brandished a toy bow and arrow. They stood back as Alvarez issued instructions to the younger colleagues who were setting up his experiment. Paul Hoch, a bearded graduate student from Berkeley, tossed one of their targets—a small melon—playfully

Luis W. Alvarez and family at the San Leandro Rifle Range on May 31, 1970. Also pictured, left to right: Sharon "Buck" Buckingham, Don Olson, Dewey Machinelli. Paul Hoch.

into the air. Another key participant was Sharon "Buck" Buckingham, a lab technician who was also an enthusiastic deer hunter.

While a third associate, Don Olson, positioned a camera, the others prepared the melons, which were wrapped in two-inch fiberglass tape to better approximate the qualities of a human skull. A wooden stand was set about thirty yards from the covered firing station in the shooting lane, where a marksman named Dewey Machinelli stood with a .30-06 rifle. The first three melons would be placed on the ground, while another four would be balanced or suspended on the target itself.

As the others withdrew, Machinelli aimed through his scope at the first melon. Alvarez watched intently to see how it would react to the bullet. Common sense dictated that it would be driven away from the shooter, but he predicted that it would move in the opposite direction—toward the source of the gunshot. It seemed paradoxical, but if he was right, it would illuminate a crucial point in the most controversial murder investigation in American history.

Unlike most of the amateur detectives who obsessed over the assassination of John F. Kennedy, Alvarez had a personal connection to the victim. Although his politics were usually on the conservative side, he had registered as a Democrat in 1960 out of enthusiasm for the dynamic young candidate—six years his junior—whom he later described as "very much a hero of mine."

Alvarez, who met President Kennedy twice, was shattered by the events of November 22, 1963. He naturally sympathized with the Warren Commission—he had served on similar committees himself—and accepted its conclusion that Lee Harvey Oswald had acted alone. Doubts would soon arise over the lone gunman theory, but Alvarez avoided being drawn into the debate for another three years.

In 1966, on the day before Thanksgiving, Alvarez listened over lunch as his graduate students argued about the current cover story in *Life* magazine, "A Matter of Reasonable Doubt." It featured the best stills ever published of the home movie taken by eyewitness Abraham Zapruder, which offered the clearest record of the crime scene in Dallas. Recognizing it at once as the kind of problem that he knew best, Alvarez reviewed the article at home, examining the pictures with typical fervor: "I got very little sleep that long holiday weekend."

Although Alvarez had no professional background in forensic analysis, he was confident in his ability to extract new information from some of the most scrutinized images of all time. For years, he had studied photos from the hydrogen bubble chamber, the massively complicated machine that he had designed to record the tracks of subatomic particles. Now he redirected those skills toward the Zapruder film, finding visual evidence in the stills to determine the timing of the shots, the frame rate, and the speed of the limousine.

It was an ingenious piece of detective work, but not every question could be resolved solely with photographs. Hoch, his graduate student, was part of an informal network of Warren Commission critics. When Alvarez later asked what he thought was the best evidence for a conspiracy, Hoch said, "The head snap." At the fatal third shot, Kennedy's head moved back and to the left, which seemed to reflect an impact from the front. This backward—or retrograde—movement was inconsistent with Oswald's position to the rear in the Texas School Book Depository, pointing instead to a shooter on the infamous grassy knoll.

If the shot came from the back, a lone gunman was still possible; if it came from the front, there must have been a second shooter—and, by definition, a conspiracy. Anyone who believed that Oswald had acted alone needed to account for the motion of the president's head. After Alvarez speculated that Kennedy had simply fallen backward when his neck muscles went limp, Hoch handed him the book *Six Seconds in Dallas* by Josiah Thompson, which made an exceptionally strong case for a conspiracy based on the physical evidence.

Although he rejected the book's conclusions, Alvarez respected its approach. In February 1969, he flew to a meeting of the American Physical Society in St. Louis, which he was attending for the first time since winning the Nobel Prize the year before. As he read Thompson's book on the plane, he was impressed by its diagrams, which led him to agree that the conspiracy theorists were right about one thing: "The president's head did not fall but was driven back by some real force."

In his hotel suite, he worked out a solution on a sheet of scrap paper. "Since I knew more physics than the [conspiracy] buffs," Alvarez said, "it didn't take me long." The answer, he felt, was surprisingly simple, once you corrected for a mistaken assumption. Most observers treated the impact as though it involved only two bodies—the bullet and the skull—in an inelastic collision, in which most of the kinetic energy was dissipated as heat. Alvarez realized that it might work differently in practice, largely because of a third factor that had to be taken into account.

Rather than transferring all its momentum to the head, a bullet would lose most of its energy in a burst of debris, as the skull's contents exploded outward. If the spray of brain matter seen in the Zapruder film was moving in the same direction as the gunshot, it could carry greater momentum than the bullet itself, requiring a motion in the opposite direction. Counterintuitively, the head would be thrust back, like a rocket ejecting its propellant. The jet effect, Alvarez concluded, could explain the backward snap without a shot from the front.

When Alvarez returned to Berkeley, he shared his work with his colleagues, one of whom recalled that the response was "tepid." His calculations were elegant, but there was no guarantee that they correctly described a phenomenon in the real world. Hoch soon convinced him that he wouldn't persuade the skeptics "unless we could demonstrate the retrograde recoil on a firing range using a reasonable facsimile of a human head as a target."

In response, Alvarez organized the series of tests that ended with his visit to the San Leandro shooting range in May 1970. As he invited his family to watch, the rifleman shot seven melons. Six rocketed back in the direction of the shooter, exactly as he had predicted. When Alvarez publicized his findings, they seemed like a scientific vindication of the Warren Commission's conclusions.

What almost no one knew was that the published results—while accurate—were only part of the story. By carefully planning the final demonstration, Alvarez obtained the outcome that he wanted, and he never mentioned a pair of earlier attempts that were much less definitive. As he later wrote in another context, "Many people think that solving a scientific puzzle is an exercise in logic that could be carried out equally well by a computer. To the contrary, a scientific detective's main stock-in-trade is his ability to decide which evidence to ignore."

AT THE PEAK OF HIS FAME, Alvarez was one of the world's most prominent scientists—a Nobel laureate whose career ranged from Hiroshima to the extinction of the dinosaurs. Today, however, he might best be known for a short scene in Christopher Nolan's *Oppenheimer.* Early in the movie, the physicist J. Robert Oppenheimer is strolling down a sidewalk in Berkeley with his lover, Jean Tatlock, when he sees a familiar figure rush out of a barber shop. Leaving Tatlock behind, Oppenheimer runs after his colleague, calling his name: "Alvarez?"

At the lab, Alvarez excitedly shows Oppenheimer and Ernest

Lawrence, his mentor, a newspaper article announcing that two German chemists have split the atom. As Alvarez dashes out of the room, hoping to reproduce the experiment, Oppenheimer heads for the blackboard, which he fills with equations to demonstrate that the fission reaction is impossible. After seeing the proof, Lawrence advises him to go next door, saying gleefully, "Alvarez did it."

In the next shot, Oppenheimer watches as Alvarez shows him a series of spiking pulses on an oscilloscope, indicating that fission is indeed taking place. Understanding its significance at once, Oppenheimer confers with Lawrence, who agrees that everyone who sees it will be thinking the same thing. Glancing uncertainly between the two other men, Alvarez asks, "What are we all thinking?" Oppenheimer responds dryly, "A bomb, Alvarez. A bomb."

Aside from a few minor departures from the facts, the scene is remarkably accurate, even if the actor Alex Wolff bears minimal resemblance to the real Alvarez, who was tall, blond, and blue-eyed. Confusion over his background would persist after his death, but the truth, as the scholar Jesús Rubén Martínez wrote, was straightforward enough: "Alvarez was a white man with a Hispanic name."

After his brief appearance in the film, Alvarez vanishes from it entirely, leaving the casual viewer without any hint of the stunning career that would follow. A 1957 profile in *Time* magazine called Alvarez the field's "prize wild-idea man," while Glenn Seaborg, a fellow Nobel winner, hailed him as "the most creative and versatile scientist I have ever known." Richard Muller, his longtime protégé, went further: "Luie Alvarez, I think, if you took a survey of physicists, could probably win the most votes for being the greatest experimental physicist of the twentieth century."

Perhaps the most revealing tribute of all came from Richard Feynman. In 1986, while serving on the commission to investigate the *Challenger* disaster, Feynman conceived of a famous demonstration—involving a piece of rubber and a glass of ice water—to show that the O-ring seals in the space shuttle's rockets became unreliable at low temperatures. At first, he was hesitant to perform it in public. "But then I think of Luis Alvarez, the physicist," Feynman recalled. "He's a guy I admire for his gutsiness and sense of humor, and I think, 'If Alvarez was on this commission, he would do it, and that's good enough for me.' "

Charismatic, physically agile, and daring, Alvarez was one of the last representatives of an era that could still see physics as a heroic enterprise. He sought new challenges in the most fascinating places imaginable, from the Great Pyramids to the Cretaceous mass extinction, which became his single greatest triumph. Reasoning brilliantly from the evidence, he concluded that the dinosaurs were doomed by an impact from space, igniting a furious debate that consumed the last years of his life.

His story is inseparable from the rise of experimental physics, which he ushered into its modern age. Even more than Oppenheimer or Lawrence, Alvarez offers the best imaginable introduction to its tools and secrets. As an experimenter, not a theorist, he thought with analogies, rather than mathematics, making his work highly accessible. While scientists in the lab rarely achieve the fame of their theoretical counterparts, the experimentalists, as Alvarez noted, have the last word on whether a theory is valid: "The only correct way to find out is to appeal to experiment."

Alvarez has been described as a scientific Indiana Jones, but his reputation as a maverick was built on a foundation of patience and discipline. "In the fields of observation," Louis Pasteur said, "chance favors only the prepared mind." To take advantage of luck, Alvarez had to remain constantly at the ready. He positioned himself to benefit from happy accidents, and many of his discoveries resulted from taking the trouble to understand a surprising result.

Physics was full of these opportunities, but only for those who could recognize them. Alvarez had a "killer instinct" for noticing what others had overlooked—what one associate called "a sort of genius for moving about and trying to see if there's one place that's an obvious bottleneck"—and refused to leave it to chance. Every week, Muller said, "Luie set aside two hours to think about what he had learned, and if they could apply to any of the unsolved problems that he knew in physics, technology, or world affairs." As Alvarez put it, "If someone thinks of a key, I can fit it into something. I have in my mind dozens of experiments which need one missing piece."

One colleague marveled that Alvarez could generate enough ideas for a hundred people: "That's what they have him there for." Another commented, "Although Luie might have one hundred ideas each day, fifty were probably useless, another twenty-five too difficult to do, and among the remaining twenty-five one or two would be worth a Nobel Prize." Muller noted that it was a deliberate strategy: "Luie had taught me that most clever ideas are proven to be wrong within a few days. If you manage to have one good idea every week, then you can expect to find one every few months that is really worth pursuing."

And none of it would have mattered if he had lacked the technical ability to realize his hunches. After mastering the arts of the machinist and glassblower, he applied himself with equal diligence to accelerator physics. He also molded himself into a relentless leader of teams, which in his hands became another tool. As the historian Peter Galison has written, Alvarez did for peacetime science what Andy Warhol did for art, turning it from a solitary pursuit into a factory. By combining the bubble chamber with the computer, he created a machine for generating serendipity.

A conventional scientist might have been content to keep mining this rich vein of material, but in an unprecedented change of heart, Alvarez turned his back on it. When the process became routine, he pivoted to problems that were unsolvable by computational power alone, as if to demonstrate the value of individual insight. Late in life, he became nothing less than the world's foremost scientific detective, which was a calling that he could pursue on his own terms.

As a result, Alvarez was one of the few experimental physicists who excelled at every stage. At a time when scientific knowledge was expanding rapidly, he resisted specialization, with what Muller called "a knack for learning just the right amount about everything." His persistence was powered by an unusual ability to live with uncertainty. He was realistic enough to prefer problems with a high potential reward, but he still relished the praise of a colleague who said, "You have a greater tolerance for risk than anyone I've ever known."

Alvarez knew that a comprehensive account of his life would double as a kind of manual on how to succeed in physics, and he labored over a book that told his side of the story. Once, at a dinner party, he fell into conversation with the wife of his rival Emilio Segrè, who was preparing a memoir of his own. When Rosa Segrè asked if he spent a lot of time reading old letters, Alvarez responded, "Oh, no, I just write down what I remember."

In 1987, the year before he died, his autobiography was published by Basic Books. *Alvarez: Adventures of a Physicist* was an engaging tour of the highlights of his career, but it was equally notable for its omissions. After his death, the astrophysicist George Greenstein mentioned him to a colleague, who replied, "Luis Alvarez—wasn't he that son of a bitch out at Berkeley?" And when Greenstein asked if Alvarez had really been "a son of a bitch," the response spoke volumes: "Well, I never knew him personally, but I understand that he was."

ALVAREZ'S COMPLICATED PERSONALITY was manifestly on display in his investigation of the Kennedy assassination. In his earliest attempts to demonstrate the jet effect, the tests at the San Leandro Rifle Range were inconclusive or worse. A variety of targets—including melons, coconuts, water jugs, and rubber balls—were shot with a rifle, but most were driven in the direction of the gunshot, or exactly the opposite of what Alvarez hoped to see.

Only the smaller melons—which bore no particular resemblance to a human skull—behaved as predicted. On the final trip in May 1970, his team shot at those targets exclusively, but the results still seemed questionable. Instead of publishing his work, Alvarez decided to shelve it. Five years later, however, the Zapruder film was shown on national television, prompting calls for a new investigation. Sensing that public opinion was shifting, Alvarez prepared a paper titled "A Physicist Examines the Kennedy Assassination Film."

In his article, Alvarez outlined the jet effect theory, but he described only the melons that reacted as he had expected. Without the full story, the tools of physics offered apparent clarity, especially when combined with his name and reputation. While it did nothing to settle the question of the shooter's location, it succeeded triumphantly in undermining the case for a shot from the grassy knoll.

And it was far from the only time that he deployed his talents in defense of an authorized narrative. He knew that anyone could suffer the same fate as Oppenheimer, who had been punished for his outspokenness by a notorious security hearing—at which Alvarez testified as a government witness—that destroyed his public career. In contrast, Alvarez gladly used his expertise and influence to shut down inconvenient lines of questioning—from UFOs to a disputed Israeli nuclear test—in the name of preserving order in an unstable world.

Alvarez's pragmatism and ambition also strained his interactions at the lab, as starkly described by the scholar Nuel Pharr Davis: "He imitated Lawrence by planning new apparatus on the biggest scale, driving construction crews of junior physicists pitilessly and showing himself off to the Berkeley public in a big red open car." For Davis, who harshly criticized Alvarez, he was "a tense and ambitious man, tired of being overshadowed by Berkeley's Nobel Prize winners and hungry for command, in which he seemed to find special release when he could use it to cause pain."

Of his behavior after the war, a colleague recalled, "Fantastic ego. Loved to tell how amazed the groups he lectured to were at how young he was. He would brag so openly you chuckled." In casual conversation, he had a habit of revealing details from his classified work, leading to multiple investigations over possible security violations. It reflected his fundamental shyness, which obliged him to aggressively assert his status, but also the unforgiving culture of the lab, where the lack of a Nobel Prize could seem like a failure—or a competitive disadvantage.

Although Davis concluded that he was viewed as "hard-headed and brutal," Alvarez could never have achieved so much without inspiring an exceptional degree of loyalty—as well as what the future Nobel laureate Saul Perlmutter called "a healthy bit of fear"—in his collaborators.

According to Richard Muller, who loved him as much as anyone, he could be a generous mentor to a select handful, but he alienated countless others: "Luie was actually hated—he was probably the most hated physicist at the Lawrence Berkeley Lab."

As Alvarez acknowledged in his memoirs, "Personalities often collide in the openly honest world of science." He could be savage, but he expected the same treatment in return, which he saw as the price of success for anyone who understood the stakes. In navigating the minefields of influence and capital, he agreed with the biologist Carl Sindermann that power was "a measure of a person's potential to get others to do what he or she wants them to do, as well as to avoid being forced by others to do what he or she doesn't want to do."

By that standard, Alvarez said, "I have been powerful for some five decades." Outside the lab, where he lacked tangible power, he learned to strategically leverage his reputation. Physics could be an instrument of persuasion, and he threw his weight behind the conclusions that he reached, even if the truth was less clear. Like many scientists who engage in politics, he was mindful of considerations that often conflicted with a disinterested appraisal of the facts.

While Alvarez was perfectly justified in arguing that Oswald was the lone assassin, his investigation of the head snap was more revealing in other ways. By cleverly framing the results, he successfully resolved the problem at hand—not the physics of the gunshot, but its public perception. Josiah Thompson, whose book had inspired the jet effect theory, grudgingly acknowledged its power to change minds: "In the whole tangled history of the case, no single individual ever played a more central role in preserving a mistaken view of the shooting."

Over the years to come, Alvarez would move even more forcefully to uphold the official account of the assassination. His objectives were obvious to the other participants in the debate, who knew exactly how formidable an opponent he could be. He recalled with satisfaction that the conspiracy buffs accused Paul Hoch of "consorting with the enemy," and one of them even asked, "We've got to stop this guy Alvarez—what have you got on him?"

I.

EXPERIMENTS

1911–1943

We cannot, in the universal imbecility, indecision and indolence of men, sufficiently congratulate ourselves on this strong and ready actor, who . . . showed us how much may be accomplished by the mere force of such virtues as all men possess in less degrees; namely, by punctuality, by personal attention, by courage and thoroughness.

—Ralph Waldo Emerson, "Napoleon"

1.

LOOKING FOR TROUBLE

1911–1936

IN 1921, A PSYCHOLOGIST NAMED LEWIS TERMAN WENT LOOKing for geniuses. Terman, a Stanford professor, wanted to recruit gifted children in California for a study that would track their development for the rest of their lives. Schoolteachers were asked to list their three best students, the youngest one currently enrolled, and the brightest child from the previous year. These candidates were given a standard intelligence test, with high scorers advancing to two additional rounds. Out of a population of 168,000 students, less than half of one percent made it into the initial group of prodigies.

One student nominated for consideration was Luis Alvarez, who was tested in the sixth grade at Madison Grammar School in San Francisco. He had been skipped ahead several years earlier, so he was the youngest child in his class—a factor that the Terman study later identified as the most reliable predictor of high scores. At age ten, Alvarez had been raised to see himself as unusually intelligent, and he had reason to expect to make the cut.

Instead, he didn't even pass the first round of testing. He did well in math, but poorly on the verbal portion, since his vocabulary wasn't especially large. His IQ would eventually be tested at 121, or well below the limit of 135 that Terman established as the cutoff. Until then, Alvarez had encountered few obstacles to his talents, and he was painfully aware that he had fallen short.

In years to come, Alvarez would often note that he was one of two future Nobel laureates who failed to join the "termites." The other was William Shockley, who led the team at Bell Labs that invented the transistor. Shockley later became known as one of the fathers of Silicon Valley, as well as for his racist beliefs about intelligence and eugenics. Alvarez wrote dryly of Shockley, "I wonder if he is ever embarrassed to remember that he 'flunked' Lewis Terman's famous test?"

Although it was disheartening to be classified below the highest achievers, Alvarez later observed that this experience was more common in adulthood. "I have not seen described anywhere the shock a talented man experiences when he finds, late in his academic life, that there are others enormously more talented than he. I have personally seen more tears shed by grown men and women over this discovery than I would have believed possible."

Alvarez was fortunate that this rude awakening occurred when he was still a child. Instead of coasting on his gifts, he looked for other traits that would enable him to excel. It led him to internalize the value of hard work and persistence—qualities that the Terman study itself would ultimately identify as more crucial to success than raw intelligence. As an adult, Alvarez would rarely encounter a challenge that he was unable to overcome, and this may have been because he was forced as a boy to consider what it really meant to be a genius.

IF HE WANTED AN EXAMPLE TO FOLLOW, Alvarez didn't need to look beyond his own family. His father, Walter, had studied medicine under Dr. Emile Schmoll, a master of the snap diagnosis. One day, for instance, a mother and her adult son entered Schmoll's office. Looking them over, he commented, "You nearly died in labor with that boy." When the mother asked how he had known, he pointed to a forceps scar over the son's eye, which testified to a difficult birth.

Walter Alvarez learned to make similar deductions using images alone. For a classroom demonstration, he was handed a chest X-ray to examine. After a moment, he said that it belonged to a woman in her fifties with several children, tall, frail, and Catholic, who suffered from polio and tuberculosis in her youth, sometimes shot pheasant, and had once been thrown by a horse.

"The thin bones and the size of the breasts showed that the film was that of a woman," Walter explained. "The fact that the breasts were hanging down suggested that she had had several children." Her long thorax and narrow rib cage were clues to her build, and he saw a Catholic medal around her neck, as well as signs of childhood illness. A fall from a horse was indicated by a healed collarbone, while three pellets of bird shot lodged in her shoulder hinted at a hunting accident.

Walter, who was born in 1884, clearly modeled himself after Schmoll—he described his mentor as "a Swiss Jew who walked in a peculiar ungainly way"—but his first hero was his own father. Luis Fernández Álvarez had emigrated from Spain, studied medicine at Stanford, and later moved his family to Hawaii, where he developed treatments for Hansen's disease, or leprosy. As a boy, Walter stood out on Oahu—like many of his relatives, he was blond and blue-eyed—and since he wasn't allowed to play with the local children, he grew close to his father instead.

Unlike his sister, Mabel, who became a celebrated painter, Walter attended the same medical school as their father, landing a position as a mining company doctor in Mexico City. Accompanying him was his new bride, the former Harriet Smyth. Harriet was born in 1885 to missionaries in China, leaving for America only as a teenager. In high school, she was grouped alphabetically with a classmate who introduced her to Walter. "If her name had been Brown," Walter said, "I'd never have known her. This just proves on what small things hang such great destinies."

At the University of California, Berkeley, Harriet trained as a grammar school teacher, although she gave up any thought of a career after following Walter to Mexico. Three years later, they returned with their daughter, Gladys, to San Francisco, where Schmoll invited Walter to join his practice. Their second child, Luis, was born on June 13, 1911. In honor of his father's colleague, Luis—pronounced "Lewis"—was given the middle name Schmoll.

A year later, the partners abruptly fell out, with an increasingly erratic Schmoll accusing Walter of "going crazy." According to

Walter, Schmoll was suffering from a psychosis of his own, which ultimately led him to be institutionalized. In the aftermath, Walter decided that his son should drop the "Schmoll." Luis grew up thinking of himself as not having a middle name, although it persisted on his birth certificate—a detail that would place him under suspicion long afterward, in a bizarre twist of fate, of spying for the Soviets.

Growing up, Luis was known as Sonny, which reflected his friendly personality. A brother, Robert, arrived a year after he was born, followed by his sister Bernice, the only survivor of a pair of twins. Although Luis was technically part Hispanic, he was hardly a person of color—his mother's side was from Ireland and New England—and he never learned Spanish, which his parents used as a secret language. When he saw a Black woman wheeling a baby carriage in the park, he innocently asked "what was wrong with her."

He became even more sheltered at four, when his parents, fearing that he had a heart condition, confined him to bed for a year. Luis soon clashed with his devout mother, who tutored him until the second grade. Anne Roe, a clinical psychologist who later used him as a case study in creativity, identified this as the source of his contrarian tendencies: "He had always disagreed with her, although not unpleasantly, was always on the opposite of everything. A bit scornful of her, actually."

After Luis entered school in 1918, he was skipped rapidly ahead. As a result, he was always younger than his classmates, which set him apart, while also encouraging his high opinion of himself. His father observed, "A child who is taught his three R's by his mother or his aunt misses many of the early democratizing influences that are to be found in a grammar school."

Walter could be quick to criticize his children, and because he was usually conducting research or treating patients, he saw them only on Sundays. At the time, he was studying the electrical behavior of the digestive system, focusing on the connection between ulcers and stress. At the Hooper Foundation, a nonprofit associated with the University of California, he built ingenious devices to film the bowels of

The Alvarez family in 1915. Standing: Walter C. Alvarez with Robert. Seated, left to right: Gladys, Bernice, Harriet, Luis.

anesthetized rabbits. On a visit, Luis saw dogs that were cut open to expose their entrails, lying in saline baths as styluses recorded their intestinal activity on rolling drums.

Luis was turned off by the messy biological side, but he was intrigued by the equipment. He began tinkering with electronics, building circuits in the family workshop, and learned to bond with his father over their shared interest in science. Walter found it easier to talk to his son while they were working on a project, so they used plans from a magazine to assemble a crystal radio set—a formative experience for many physicists, including Richard Feynman. Luis later regretted that he didn't follow up by becoming a ham radio operator, which would have given him skills that turned out to be tremendously useful.

When Luis was twelve, his father, who loved the outdoors, took him on a hike in the Sierra Nevada. Fifty miles from town, Luis felt stomach pains, and Walter told him to lie by the fire. After an hour, he felt better, and it was only then that he realized that his father had been calmly sterilizing knives in boiling water, in case he had to remove his son's appendix in the wilderness.

Later, at his first Boy Scout summer camp, Luis cut his index finger on a pocketknife. It throbbed and turned yellow, but he ignored the pain. On arriving home, he showed it to his father and grandfather, who were horrified. His veins were dark blue, and when he removed his shirt, they saw that his lymph nodes were swollen. If he had waited even one day longer to start treatment, he might have died, leading him to marvel afterward at how close he came to "losing my precious ass."

Given his son's interest in science, Walter decided to send him to Polytechnic High School in San Francisco, which was geared toward vocational training. Luis felt awkward around his classmates, who mostly came from less privileged families, and withdrew into academics. Math was easy—another student admonished him for taking an exam with a fountain pen—and he could grasp geometrical proofs at a glance. He often sketched alone in Golden Gate Park, although the only time this skill came in handy, he said much later, was at the atomic bomb test in Alamogordo.

Luis often wondered how his personality would have suffered if he had stayed at Polytechnic. Fortunately, halfway through his second year, his father landed a research position at the famed Mayo Clinic. In 1926, they moved to Rochester, Minnesota, where Luis thrived in the more studious atmosphere. For the first time, he went on dates, usually with freshman girls who were his own age, and since he was less ahead of his peers than before, he looked for other ways to stand out.

He was fascinated by the book *Creative Chemistry* by Edwin E. Slosson, who wrote, "In the eyes of the chemist the Great War was essentially a series of explosive reactions resulting in the liberation of nitrogen." After hearing a lecture on electromagnetism, however, he grew more interested in physics. Walter paid for weekend lessons with a machinist at the Mayo Clinic, where Luis learned the art of gear cutting. In their garage, he built a model Spanish galleon—turning dowels on a lathe for the masts, weaving the rigging—and displayed a gift for methodical work.

All the while, he was secretly testing his tolerance for risk—the shadow side of his patience and attention to detail. At night, he and a friend would sneak past the guards at a construction site to climb inside a smokestack, search the powerhouse, and clamber over the beams of an unfinished tower. Curious people, he concluded, had the right to exhibit "a controlled disrespect for authority."

He expected to go to UC Berkeley, which his mother and sister Gladys had attended, but his teachers recommended the University of Chicago instead. At the time, it was associated with all three American Nobel laureates in physics—Albert Michelson, Robert Millikan, and Arthur Compton—and Berkeley had yet to reach the same level. Within a few years, their positions would be reversed, but for now, he followed the advice of his elders.

In 1928, after graduating fifth in his class, he joined his brother, Bob, on a Sierra Club trip to the Tonquin Valley in Alberta, Canada. One day, they climbed a glacier, following a steep path that lay a hundred yards from a drop of fifty feet. They made good progress across the packed snow, but when Luis glanced down, he saw pockmarks in the ice. He belatedly recognized them as scars from falling rocks, which meant that he had wandered into an avalanche chute.

As this thought crossed his mind, he saw a boulder crash, rebound, and tumble their way. Yelling at Bob, he dashed across the slope to safety, but slipped on a patch of ice. As he slid helplessly toward the edge, he remembered his climbing manuals. Taking the handle of his ice axe in his hand, he rolled onto his stomach and buried the pick in the slope, which brought him to a stop.

Back in Rochester, when Luis checked his mountaineering books, he realized that he had done exactly the wrong thing. When jabbing the axe into the ice, climbers were supposed to hold the head, not the handle, which could slip through their grasp, leaving them with no way to arrest their fall. Reading this, he broke into a sweat. By following his intuition, he had come close to killing himself.

He was reminded that books were no substitute for experience, especially if you forgot what you had read. It was the second time, after his infected finger, that he had been saved from death by sheer luck. As he set off for college, all he knew was that he wanted to succeed, even if he wasn't entirely sure why, and that the only way to live with chance was to be prepared for it when it came.

LUIS ALVAREZ ARRIVED at the University of Chicago on a bright autumn afternoon in September 1928. As he and his father walked admiringly across the campus, they ended up at the physics department in Ryerson Laboratory, where the only person in sight was a professor installing a crystal spectrometer. It was Arthur Compton, who had won the Nobel Prize ten months earlier for demonstrating that light could be treated as a stream of particles. Compton struck Alvarez as old—in fact, he was just thirty-six—but he seemed happy to talk with the two visitors.

It seemed like an auspicious beginning to college life, although Alvarez wouldn't set foot in Ryerson again for over a year. For now, he had only one objective in mind—he wanted to major in chemistry. His retreat from his interest in physics was guided by practical considerations. Most laymen didn't even know what a physicist was, and for years afterward, whenever people asked what he did, he found it easier to tell them that he was a chemist.

Alvarez had been raised in a culture shaped by what he later called "a chemist's war." During his first quarter, however, he earned just a B average, while a course in German seemed likely to end in a D. He was saved when a flu outbreak resulted in the cancellation of all classes, leading the instructor to give him an automatic B—the only reason that he later made Phi Beta Kappa.

"I learned that luck can be as important to success as skill," Alvarez remembered, but he knew that he wouldn't escape so lightly again. The following year, he took freshman physics. Memorizing material from textbooks was dull, but he was even more repelled by chemistry. He hated the hydrogen sulfide fumes and the stink of benzene, and he showed no aptitude for the lab work: "I never produced anything in my test tubes except a succession of black, gooey residues."

At the end of his sophomore year, his grade for one course was based largely on a single analysis of an unknown substance. After he lost part of the sample, he adjusted his answer by an arbitrary amount, which turned out to be exactly right—although he still earned only a B overall. "I had two reactions to this experience," Alvarez wrote. "First, I couldn't stand the idea of being a B chemist . . . when it was beginning to be clear that I might be an A physicist. Second, I imagined with horror a lifetime in chemistry always guessing the answer."

Alvarez's academic advisor was the astrophysicist Henry Gordon Gale, who later became the dean of physical sciences. In the fall of 1930, Gale steered him toward a course taught by George Monk, an expert in optics, on the experimental physics of light. The centerpiece of Monk's lab in the Ryerson attic was an optical spectrometer. When light struck a metal diffraction grating etched with thousands of grooves, it was dispersed into a spectrum of colors. As Alvarez took his first measurement—the wavelength of a mercury lamp—the elegant equipment evoked the technological exhibitions and construction sets that he had loved as a boy.

Alvarez credited the class with changing his life. He was especially fascinated by the diffraction gratings, which were produced in the Ryerson basement. An automated diamond stylus took five seconds to inscribe each of one hundred thousand rulings, traveling three miles by the time it was done. A broken tip or power outage could ruin days of work, and the temperature had to be precisely controlled, since a tiny fluctuation could distort the pitch of the screw that spaced the parallel lines.

Another intriguing machine was the interferometer, which used mirrors to merge two light paths into an interference pattern that allowed for fine measurements of the beams. It had to be carefully adjusted, so the students turned it into a game. Armed with a stopwatch and a stack of dimes, they twisted the screws until the characteristic colored fringes appeared. A good time was fifteen seconds, Alvarez recalled, adding wistfully, "Now, it would probably take me ten minutes."

When he switched to physics, Monk advised him to skip the

textbooks in favor of Albert Michelson's original papers. To Alvarez's delight, he understood the primary sources: "The physics library was so engrossing that I had to force myself to leave it for food or friends." Reviewing vintage journals became a lifelong habit, aided by his exceptional memory. He could remember the exact spot on a page where he had once seen an illustration, and he was able to find almost any article that he had ever read within minutes of entering the library stacks.

His decision to change fields was confirmed by a course on electricity and magnetism in the spring of 1931. The instructor was J. Barton Hoag, whom Alvarez later praised as his most inspiring teacher. Until then, he had taken for granted that his lab equipment was properly maintained. Reality, Hoag warned, was more insidious, and he emphasized the importance of questioning one's tools.

To underline this point, Hoag encouraged students to put "trouble" in one another's experiments. Given a chance to indulge his prankish side, Alvarez played devious tricks. When a classmate had to take magnetic field measurements, for example, he found a way to lower the readings by twenty percent. "Hoag rewarded me with an A," Alvarez wrote, "remarking that even though he thought my trouble was very clever, I would find that nature could be much more perverse."

When Alvarez decided to build a telescope, using a book for hobbyists, Hoag set him up in the basement. To make the mirror, he had to walk in a circle for hours, pushing one glass blank against another. On a visit, Gale was appalled: "Why don't you come upstairs and use the real machines?" He brought him to the optical shop, which Michelson—who died before Alvarez could meet him—had filled with his equipment and assistants. The craftsmen adopted Alvarez as their apprentice, and although he struck them at first as "a brash kid," he impressed them with his diligence.

Using a grinding machine, he needed only a day to wear the glass down to the right dimensions. Next came weeks of polishing by hand, until the surface was correct to within a few millionths of an inch. Gale gave him a key to the student shop to build the mounting—Alvarez claimed that he was the youngest person to ever receive such a privilege—and the finished telescope was installed on the roof. In the end, he barely used it, but the project heightened his confidence.

His first truly original idea came to him in Rochester in the summer of 1931. When a phonograph record was struck by light at a shallow angle, he noticed a faint rainbow on its surface, since the grooves served as a crude diffraction grating. It wasn't good enough for a real spectrometer, but he arrived at a clever way to measure the wavelength of any color.

Alvarez conducted the experiment in his family's living room. On the wall, he placed a light bulb and a sheet of paper with horizontal markings. He broke off a piece of the record perpendicular to the grooves, about two inches wide, so that he could sight along it with one eye. Standing at a fixed distance from the wall, he held it up, looking across its surface at the reflected rainbow. Then he selected a color and estimated how far it seemed from the light source, using the ruled lines on the paper, as though it were actually on the wall itself.

For the next step, to find the distance between the grooves, he counted the revolutions that it took for the phonograph needle to travel one centimeter. With this information, he could use basic trigonometry to calculate the wavelength. Back on campus, Gale told him to write it up as an educational tool. It had little practical significance, but it vividly displayed his ingenuity.

In the fall, Hoag proposed that he build a Geiger counter. Although the literature was vague on the details, the principle seemed straightforward enough. A high voltage was applied to a tube that conducted a charge whenever it encountered a radioactive particle, producing a pulse that was amplified into an audible click. At the time, it was a new invention, so nobody in Chicago had even seen one.

"The project was both important and at the limit of my skills," Alvarez remembered. Hoag clearly saw it as a test of his potential, as well as his ability to learn new crafts, like glassblowing, while dealing with limited resources. Most of the electrical components at Ryerson could fit into a single cigar box, so he had to buy much of the equipment himself. Hoag provided a power supply that yielded five thousand volts, but Alvarez was terrified of it, discharging it with a copper bar on a long pole before he dared to go any closer.

His power supply became a source of "trouble" as devilish as anything from Hoag's classes. It had been built by another student, who hadn't bothered to label the positive and negative terminals. Alvarez tried to tell them apart by visual inspection, but he later found that he had hooked it up backward, which he didn't realize for months. "This misapprehension did not keep me from getting my Geiger counter operating, for reasons that I still don't understand."

Once it was working, he impressed his friends by showing how it clicked near the radium dial of his watch. People who could build Geiger counters were seen as "wizards or sorcerers," and Alvarez later recounted a joke by J. Robert Oppenheimer: "One school [of thought] held firmly that the final step before one sealed off the Geiger tube was to peel a banana and wave the skin three times sharply to the left. The other school was equally confident that success would follow if one waved the banana peel twice to the left and then once smartly to the right."

Alvarez thought that it was the worst counter ever built—it suffered from background noise even when it was nowhere near any radiation—but he was invited to give a talk on it. One attendee was Arthur Compton, who advised using the "coincidence technique," in which two counters clicked only when both generated a pulse simultaneously. This would reduce the background problem, while also allowing it to serve as a telescope for observing cosmic rays, since a pair of counters could determine the direction of a particle from space. Seizing the chance to work with one of the country's most famous physicists, Alvarez agreed to build a version that could be deployed in the field.

As Alvarez neared the end of his undergraduate years, he had reason to see them as a success. On the social side, he had pledged Phi Gamma Delta soon after arriving in Chicago. To his housemates, he was known as "the mad Spaniard from Minnesota" or "the Spanish Swede," although he usually went simply as Luie.

He was largely untouched by the economic depression that followed the stock market crash of 1929, and he was equally oblivious to social issues. In Monk's lab, his partner had been one of the department's only Black students, but although they reportedly chose each other out of "unspoken consent," Alvarez said nothing about him in his memoirs. He witnessed racial disparities on a trip to the South with his brother, as well as in Chicago, but he conceded of the latter, "It did not occur to me that [the people] were victims of discrimination and injustice."

As an undergraduate, he had plenty of distractions. On days off, he snuck through the stage doors of theaters with his fraternity brothers, confident that the ushers wouldn't stop them if they acted like they belonged. Prohibition was in effect, so he made a connection with a bootlegger, although the merchandise left him sickened ever afterward by the taste of gin. Another source of spirits was the alcohol used to sterilize equipment, which students signed out at their discretion: "My scientific friends exhibited extraordinary anxiety about the cleanliness of their laboratory glassware."

Alvarez frequently played billiards, which was popular with the physics crowd, but he was less interested in team sports, since he preferred to show off his individual talent. He followed the example of his father—whom he had recently honored by taking "Walter" as his middle name—by pursuing gymnastics. Before long, he could do giant swings on the horizontal bar, although he was so tall that he had to push apart the mats to get a few extra inches of clearance. He spent ten hours a week at the gym, but it never occurred to him to invent new routines, which would have been second nature to him later in life.

Like billiards, which was a game of collisions and dynamics, gymnastics had obvious affinities to physics. In retrospect, however, he was surprised that he never thought of it in terms of centrifugal forces, elastic limits, or torque. He had always been drawn to challenges that depended on the interaction of the body with its surroundings, but for now, physics was confined to the lab, rather than a tool for thinking about the world. "Intellectually," Alvarez concluded, "I was a different person then from the person I later became."

❦

IN THE SPRING OF 1932, Alvarez started graduate school, concentrating on the cosmic ray telescope that he had promised to Compton. The study of cosmic rays had begun decades earlier, after scientists noticed that shielded electroscopes were discharged by a form of penetrating radiation that evidently came from outer space. More recently, Compton had found that the rays were more intense near the poles than the equator, indicating that they were charged particles.

Because of the dynamics of the earth's core, magnetic field lines radiated from the South Pole toward the North Pole, like iron filings around a bar magnet. A charged cosmic ray could reach the lower atmosphere more easily if it moved with the lines of force, rather than against them. At high latitudes, the magnetic field was aligned with incoming particles, so they were more likely to be detected there than at the equator, where the field lines were perpendicular to their approach.

Charged particles would also swerve one way or the other based on "the right-hand rule." If you pointed your right index finger along the path of a positively charged cosmic ray, then turned your hand to lift the other fingers in the direction of the magnetic field, the force on the particle would be indicated by your thumb. Negative particles would be pushed the opposite way. If the rays were positively charged, a Geiger counter telescope should detect more rays when pointed west, while negative particles would be more intense from the east. The consensus was that they were negatively charged electrons, but no one knew for sure.

Over the Thanksgiving holiday in 1932, Alvarez was invited to a meeting with Manuel Sandoval Vallarta, a physicist from the Massachusetts Institute of Technology, who thought that scientists were looking in the wrong place. If you were too far north, charged particles would race down the field lines as they dipped toward the pole, leaving no measurable difference from either direction; but if you were too far south, they would be deflected away before they even reached

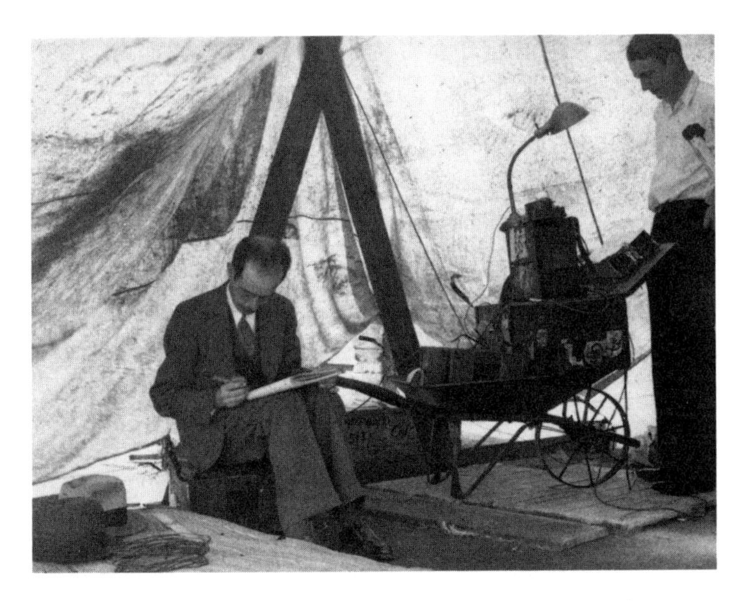

Manuel Vallarta and Alvarez with rooftop cosmic ray telescope in Mexico City in April 1933. Lawrence Berkeley National Laboratory.

the lower atmosphere. Only in a narrow intermediate zone would the effect be noticeable.

Vallarta concluded that Mexico City, his birthplace, was at a latitude and altitude that was favorable for observations. Another physicist, Tom Johnson of the Franklin Institute, agreed to make the trip, and Alvarez was urged to join as well. While he hardly needed the encouragement—it was a perfect showcase for his abilities—he had to hustle to be ready on time. Modifying his equipment to run on batteries, he tested it in a Chicago park before boarding a train to Mexico.

On their arrival, the two researchers hoisted their telescopes to the roof of the Hotel Geneve, where Alvarez pitched a canvas tent to protect his equipment. He had mounted it on the hinged lid of the wooden battery box, planning to angle it east or west as needed, but Johnson pointed out that this could produce a spurious effect if the counters weren't exactly centered. Embarrassed at overlooking a possible source of error, Alvarez bought a wheelbarrow, hauled it up to the roof with a rope, and used it to rotate his telescope every half hour.

After a few days, they confirmed that a higher count was coming from the west, indicating that the primary charge of cosmic rays was positive—the opposite of what most physicists had predicted. At a celebratory dinner, Alvarez listened attentively as Johnson shared stories about his onetime Yale tennis partner Ernest O. Lawrence, the inventor of the particle accelerator known as the cyclotron, who had studied at Chicago before finally settling at Berkeley.

On his return, Alvarez wrote a paper with Compton, who put the younger man's name first. He was pleased that his advisor was willing to leave him alone—Compton visited Alvarez's research room exactly once—and trusted him to come up with worthwhile ideas himself. Alvarez more than lived up to his expectations, prompting Compton to describe him later as "the most brilliant student he ever had."

As Alvarez dreamed up experiments, he remained irritated by the lack of resources on campus. All of the laboratory clamps in the shop had been stolen by other students, who set up a black market in equipment, like a miniature lesson in the economics of the Great Depression. Essential goods were already scarce, and they were further diminished by hoarding.

Approaching it as another game to play, he decided to challenge the "clamp cartel." One Sunday, when the lab was empty, he methodically searched for hiding places. Noticing an air duct in a research room that belonged to a cartel leader, he climbed onto a desk to look. To his delight, he saw scores of clamps lined up on the duct, and similar stashes turned up elsewhere.

Alvarez hid them in a secret location of his own. The following morning, he casually mentioned his discovery to two cartel members. Going next door to eavesdrop, he heard them wonder if he had found the stash in the grating room. He ran to get the clamps, giving him an undisputed monopoly, and later chaired a meeting where students claimed one clamp at a time until all were taken.

While working as a lab assistant, Alvarez played a prank on an instructor he disliked. In a shaft by the elevator hung a Foucault pendulum—a heavy weight on a long wire that slowly traced a clockwise circle, demonstrating the rotation of the earth. For lectures, the instructor would gather students at the pendulum, mark the plane of its swing, and return an hour later to show that it was moving as expected.

Alvarez wanted to make it look as though the pendulum had rotated in the wrong direction. After the instructor and students had departed, he stopped the ball and tied it to the wall with string, ten degrees counterclockwise from its previous position. He waited until the class was almost over, then burned the string to set the ball in motion again. When the instructor returned, he was baffled to discover that the pendulum had traveled backward.

Of all the graduate students, Harold Plumley, who was seen as the brightest one there, was the one to whom Alvarez was closest. They once made a bet about the size of the spark that would appear if they pulled the main power switch at Ryerson. On a night when a diffraction grating wasn't being ruled, they snuck in, seized the handle together, and yanked it down. Alvarez was disappointed: "There wasn't much of a flash and no noise. The lights simply went out." Turning it back on, they quietly left.

The next morning, Dean Gale was furious about the incident. As a crowd of students gathered, Alvarez and Plumley confessed, emphasizing that they had been careful not to ruin a grating. Fortunately, Gale was amused—he agreed to let them off if they told him how large the spark had been—and informed the onlookers that he had nearly expelled Lawrence for a similar offense. "Do you know what that scoundrel did? He had a master key that he filed down himself."

Everyone exchanged glances. Master keys were officially reserved for staff, but every graduate student had secretly made one by filing down a key to his research room, which was the only way to enter the building after hours. Alvarez later observed that there was an apparent correlation between almost being expelled by Gale and winning the Nobel Prize.

His attention was soon drawn to the Chicago World's Fair of 1933. Alfred Sloan, the president of General Motors, wanted something impressive to open his company's exhibition. At Compton's suggestion, Alvarez built a setup that would use a cosmic ray pulse to start a production line. There wasn't time to test it, so he discreetly ran a pair of wires to a junction box, ready to utilize it as a backup switch if the equipment failed. Fortunately, it worked.

Alvarez, who rarely hesitated to use physics for a publicity stunt, later became involved in another showpiece at the fair that memorably combined science with physical daring. A navy balloonist, T. G. W. "Tex" Settle, asked him to assist in measuring cosmic rays at a record elevation. They would ascend in a spherical metal gondola—Alvarez gave himself a nasty shock when he unthinkingly picked up an electrical lead while installing their gear—to a height of over ten miles. Alvarez was eagerly preparing for parachute training when he learned that the plan had changed at the last minute, and Settle would go alone.

He was disappointed, but he still looked forward to the flight on August 5, 1933. At Soldier Field, the balloon was launched at three in the morning, allowing the hydrogen to be warmed by the rising sun. As Alvarez watched, clouds of vapor emerged as Settle adjusted his altitude. Suddenly, he began to plummet. Jumping into his car, Alvarez raced to a railway yard a mile to the west, where the balloon had landed. Settle, unhurt, explained that his valve cord had become tangled. They guarded the balloon from onlookers—many were smoking, unaware of the risk of an explosion, and a few tried to cut pieces of the bag as souvenirs—until the police arrived.

Three months later, the balloon made a successful ascent, with Alvarez speaking to Settle by radio from the ground. He remained fascinated by flight, which was another test of physics in the real world. After his grandfather sent him fifty dollars for Christmas, he spent it on flying lessons at Midway Airport. Although he could afford only three hours of training, he managed to pull off his first solo landing, outfitted with a helmet and white scarf.

His most significant encounter that year occurred elsewhere. Over the summer, Alvarez had attended a meeting of the American Association for the Advancement of Science, which featured a talk about the cyclotron by Ernest Lawrence. During a coffee break, Alvarez introduced himself, knowing that they had a personal connection—his older sister was working as Lawrence's secretary. He stuck out his hand. "I'm Gladys Archibald's brother."

"Yes, I've heard about you," Lawrence said. He was also familiar with Alvarez's father, who served on the board of the Macy Foundation, a nonprofit devoted to medical research. After Gladys told Walter about Lawrence's work with radioactivity in the life sciences, he had supported a grant for the Berkeley lab. It impressed Lawrence as a notable act of generosity, so he decided to return the favor.

At the end of the summer, Lawrence found himself in town for a day between trains. Calling Alvarez from the railway station, he proposed that they tour the fair. Alvarez was excited by the invitation, which came from exactly the role model he needed. At thirty-two, Lawrence was a decade older, but they were physically and temperamentally alike. Both were lean, over six feet tall, with glasses and sharp blue eyes, and both hid their ambition behind an air of affability.

Lawrence's charming smile concealed his quick temper, his readiness to drive others to their limits, and the bouts of depression that he overcame with work. For now, however, he struck Alvarez as approachable and relaxed. At the fair, they walked the Streets of Paris Exhibition, saw a fan dance by the famed burlesque artist Sally Rand, and listened to an outdoor orchestra. Alvarez was gratified that other students seemed envious of his proximity to the great man, and back at the hotel, Lawrence cheerfully said, "Let's have a nightcap."

The meeting occurred at a point when Alvarez was conscious of how far he still had to go. He was working on a difficult experiment to measure how free electrons, or beta rays, were deflected within magnetized iron. Borrowing glass radon needles from a radiologist, he carried them in a lead box on the train to the lab, where he carefully unpacked them with tweezers. It gave him valuable experience with handling radioactive materials, but after writing up the project as his master's thesis, he realized that he had made a serious mistake in the analysis.

Another project was inspired by a discrepancy in measurements of the charge of the electron. One value had been obtained, in a celebrated experiment with charged oil droplets, by Robert Millikan; another used a diffraction grating to measure an X-ray line. The results were different, so one of them had to be wrong. Alvarez suspected that there was an issue with the diffraction approach. Since everyone assumed that previous calculations were correct, any errors in the published equations might have gone unnoticed for decades.

Alvarez decided to check the standard grating equation by precisely measuring a line of known length and comparing it to the calculated value. He expected that it would take just a few months, but it turned out to be much more complicated. To produce a blue line, he built a mercury vapor lamp, which often blew up. Mercury droplets sprayed across his unventilated research room, and he later felt relieved that he had avoided permanent brain damage.

Taking accurate measurements also demanded that he scale up the setup enormously. While a standard spectrometer could fit on a table, he wanted the distance from the light source to be ten times longer. For his first attempt, he placed the lamp and photographic plates at opposite ends of a hallway, aligning it all with a surveyor's level, but air currents caused the images to blur. Using fans to generate a regular airflow, he finally obtained a good measurement—only to find that it was identical to the expected value. The diffraction method was fine.

It later turned out that Millikan's oil drop experiment had been flawed, but for now, it felt like a waste of effort. Compton, unfazed, told Alvarez to write it up as his doctoral thesis. No one cared about the subject, Compton said, as long as the researcher did interesting work elsewhere. After two consecutive dead ends, however, Alvarez worried about how he was spending his time. While he clearly had patience and skill, he lacked what physicists called "taste," or the gift of identifying projects that were both achievable and important.

This required a suitable mentor, and Alvarez already had one in mind. In the summer of 1934, he had visited Lawrence in Berkeley. The Radiation Laboratory, which housed the cyclotron, occupied an unassuming clapboard building near the physics department in LeConte Hall. There were no internal doors, and everyone freely shared their equipment, as well as candid feedback on experiments. The culture there provided a stark contrast to Ryerson, where it was considered bad manners to criticize others. Physics in Chicago, he thought, was slowly dying, even as everyone reminisced about the good old days, "like a Tennessee Williams play."

Over dinner with his parents at Lawrence's house, Alvarez was charmed by the physicist's wife, Molly, who gave up a career in bacteriology to support her husband's work. He was left with the impression that Lawrence—to whom he wrote a letter praising "the enthusiasm which I liked so much at Berkeley"—was just getting started. Lawrence had funded and built the most impressive piece of physics equipment in the country before he even knew how it could be used. Money and talent would continue to flow west, as if directed by invisible lines of force, to the lab that Alvarez called "the most exciting place I had ever seen."

Alvarez was also preoccupied with developments in his personal life. For two years, he had dated the younger sister of a fraternity brother's fiancée, but they broke up shortly before he entered graduate school. They had occasionally gone on double dates with another friend and an undergraduate named Geraldine Smithwick. Late in 1933, Alvarez began taking Geraldine to dances, and before long, he suspected that this pretty, outgoing senior was the woman he would marry.

Geraldine, or Gerry, was born in 1913 in Omaha, Nebraska. She adored her father, Jeremiah, who overcame his limited education—he took night school classes in accounting—to become the comptroller of the Chicago meatpacker Swift & Company. In college, she loved theater and student government, and she was one of just twenty freshmen selected for a famous seminar on the great books taught by the university president, Robert Maynard Hutchins, and the philosopher Mortimer Adler. The future publisher of the *Washington Post*, Katharine Graham, who took the course later, recalled, "The methods they used often taught you most about bullying *back*."

It was good training for life with Alvarez, who proposed in the fall of 1935. As Geraldine typed his thesis, Alvarez studied for his orals and thought about their future together, knowing that he didn't want to be supported by his wife. Geraldine worked as the secretary for the president of a charitable foundation, the Rosenwald Fund, which left her with a lasting commitment to social justice—a subject to which Alvarez was largely indifferent. As for his own prospects, most physicists at the time earned a living by teaching or prospecting for oil with seismic techniques in Texas. Neither was especially alluring, so Alvarez decided to approach the scientist who had impressed him the most.

He had seen Lawrence once more at a convention in Washington, DC, where the older man smiled at him from across the room. Alvarez followed up with a proposal to study beta rays in Berkeley, which Lawrence wrote was "a splendid idea." The visit never occurred, and Lawrence was unable to secure him a fellowship. Hoping to get a job anyway, Alvarez asked Gladys to put in a good word on his behalf. On April 6, 1936, he received a thrilling telegram:

> Lawrence says come on out. If you get no fellowship can pay you probably a thousand a year depending on how much money he can get for budget. Can't start pay until July, but plenty of work any time you come. He sends congratulations for wedding. My letter follows. Gladys.

Eagerly accepting the offer, which would oblige Geraldine to leave her family and career, he prepared to move to California.

"The reason that I'm not wearing cowboy boots now," Alvarez later said, "is only that my sister was Ernest Lawrence's part-time secretary." He conceded that other considerations may have influenced Lawrence's decision: "Lawrence, who was a very practical man, probably said to himself, 'Alvarez's father isn't rich, but he does have his hand on a lot of foundation money that could flow our way if his son could reinforce his enthusiasm for our work.'"

On April 15, 1936, a few days after his graduation, Alvarez and Geraldine were married in a university chapel. They immediately embarked in their Model A Ford for Berkeley, where they moved into his sister's apartment in May. When Alvarez turned up at the

Radiation Laboratory, the door was opened by a graduate student named Ernest Lyman. He brought Alvarez to John Livingood, who was in charge while Lawrence was at a conference. When Livingood asked when he could start, Alvarez replied, "As soon as I can get my coat off."

2.

COMING OF AGE IN BERKELEY

1936–1939

OF HIS EARLY DAYS WITH LAWRENCE, ALVAREZ RECALLED, "All he expected from me I think was lots and lots of hard work, unlimited hours, repairing the cyclotron and operating it, and getting in a little physics if there was any time left over." At first, there was no apparent reason to anticipate more. Alvarez seemed content to help out as needed, rather than jumping into his own research: "I just felt that I ought to put some money in the bank before I started withdrawing it."

On a practical level, Alvarez understood that he would benefit enormously by studying those around him, particularly Lawrence himself. Thanks to access and good timing, he had been given the chance to participate in one of the most exciting eras in the history of physics. Alvarez wasn't quite present at the creation, but he was close, and he could think of no better way to spend his first year than by observing the man who had made it all possible.

Ernest Lawrence, who was born in South Dakota in 1901, always thought of himself as "a country boy." After doing exceptional work at Chicago and Yale, he was recruited by Berkeley during a fundamental shift in the field. "From the discovery of radioactivity in 1896 until 1919," Alvarez noted, "nuclear physics was essentially an observational science." Physicists, like astronomers, had been limited to looking at nature, rather than controlling it.

The situation changed radically after Ernest Rutherford described the transmutation of an element in the lab. With the aid of Patrick

Blackett—the Cambridge professor whom the young J. Robert Oppenheimer famously tried to poison with an apple laced with cyanide—he found that nitrogen transformed into oxygen when struck by an alpha particle, a free helium nucleus with two protons and two neutrons. It turned nuclear physics into an experimental science, allowing researchers to actively explore the atom, instead of watching it passively.

Lord Rutherford became one of Alvarez's heroes, but his approach—which used radium as a source of radioactivity—was slow and tedious. Before long, physicists turned to generating a beam of charged particles, which would serve as a more powerful probe. A race ensued to accelerate protons to a kinetic energy of one million electron volts, or MeV, which was thought to be necessary to disintegrate one element into another. One of the leading contenders was Lawrence, who spent his evenings looking for clues at the library.

In 1929, Lawrence saw a paper in German by the physicist Rolf Widerøe. Although he was unable to read the text, he was riveted by its illustrations of a linear accelerator, which took advantage of the force exerted by an electric field on charged particles. Two cylindrical electrodes were set in line to make a tube; potassium ions were accelerated at one end by an alternating current; and the particles received a second kick at a gap in the middle.

If the gap coincided with the right point in the cycle, the ions would exit with twice as much energy as before. By lining up more tubes, which had to be successively longer to account for the rise in speed, particles could be accelerated repeatedly, but this struck Lawrence as impractical: "Simple calculations showed that the accelerator tube would be some meters in length, which at that time seemed rather awkwardly long for laboratory purposes."

With the aid of an invaluable colleague, Milton Stanley Livingston, Lawrence wound the accelerator path into a spiral. A cyclotron was shaped like a doughnut, with a pair of curved electrodes between the poles of a magnet. When a source of protons—like ionized hydrogen—was placed at the center, the particles curved in the magnetic field, repeatedly crossing the same gap at a constant rate. Even if the accelerating voltage was relatively low, their energy would rise with every revolution until they emerged in a beam powerful enough "to bombard and break up atoms."

At first, Lawrence saw it as a tool for exploring the nucleus, but he drastically revised his plans in 1934, after Irène and Frédéric Joliot-Curie used alpha particles to turn aluminum into radioactive phosphorus. The advent of artificial radioactivity—the transmutation of a stable element into an unstable isotope that emitted radiation as it decayed—led to a "gold rush" at Berkeley, as the cyclotron lab transmuted as many elements as possible. Lawrence hoped that radioisotopes could treat cancer, but every previously unknown isotope was also a publishable discovery in itself.

This was the prevailing atmosphere at the Radiation Laboratory when Alvarez arrived in May 1936. As Ernest Lyman took him on a tour, he looked carefully around the building. Past a workshop and control desk was the cyclotron itself, followed by a room with a linear accelerator and hundreds of caged mice. The mice were the property of Lawrence's younger brother, John, a physician who was studying the effects of radioactivity. Alvarez was assured that he would soon get used to the smell.

They returned to the main room to take a closer look at the twenty-seven-inch cyclotron. Its diameter referred to the size of the "can," the cylindrical vacuum chamber in which accelerated particles traveled their spiral path, but the machine as a whole was much larger. The chamber was wheeled into place between the upper and lower poles of a gigantic electromagnet, like two huge hamburger buns, which were hooked up to a bewildering array of equipment.

Lyman put Alvarez to work at once. A transformer on the cyclotron had burned out, and they couldn't afford to replace it. Lifting it away, they emptied out the insulating oil and set it on a workbench, where Alvarez helped with the repairs, knowing that he would need to master these skills himself. When he went home for lunch with Geraldine, he was stinking of the oil, which would cling to him—instantly marking him as a member of the lab—throughout his time in Berkeley.

M. Stanley Livingston and Ernest O. Lawrence at the twenty-seven-inch cyclotron in Berkeley in 1934. National Archives.

As Alvarez commenced his training, he was introduced to Ryōkichi Sagane, a young researcher who planned to build a cyclotron in Tokyo. Although they worked closely together over the next two years, Alvarez never learned much about Sagane's private life. If he had asked, he would have found that they had similar backgrounds, starting with an intimidating father—Sagane was the son of Hantaro Nagaoka, one of the most celebrated physicists in Japan.

For now, all they shared was their lowly status as they were initiated into the cyclotron's mysteries. When the machine was started up each day, the beam was barely detectable. To increase it, the operator pushed a slide that raised or lowered the magnetic field to match the electric current. Because of the lag time before any visible change, it was more an art than a science, and Lawrence was still the best, followed closely by his wife, Molly.

To protect the beam from air resistance, the vacuum chamber was coated with beeswax and resin, along with patches of red sealing wax. Because of the enormous stresses from the magnet, it constantly sprung leaks, which were sealed by briefly remelting the wax with a flame. More elusive cracks were found by passing natural gas over the exterior, watching a gauge for a slight rise of conductivity inside the chamber. Detecting these seepages often took Sagane and Alvarez—who stubbornly tracked down a fault line that became known as "Luie's leak"—an entire day.

By modern standards, the equipment was crude. The cyclotron had to be turned up slowly, which required a control knob. As an inexpensive solution, the crew connected each transformer to a glass column of water, with a graphite electrode suspended inside. By lowering the electrode, they reduced the amount of water between it and the power source, gradually increasing the current. It was controlled by a rope and pulley connected to a steering wheel, like something from a Rube Goldberg cartoon.

The frequent breakdowns and accidents—the magnet could grab a wrench from a technician's hand and send it crashing through a chamber window—affected the research that was pursued. Since researchers couldn't rely on long periods of uninterrupted operation, they needed projects that would yield experimental data even when the cyclotron wasn't running.

Their answer was to turn the lab into what Alvarez described as "a radioactivity factory." A target element would be placed in the cyclotron's bombardment chamber, exposing it to a beam of deuterons—nuclei of "heavy hydrogen," with one proton and one neutron—to make it radioactive. The result of these collisions, ideally, was a previously unknown isotope. Most researchers signed up for bombardments right away, reserving part of the periodic table for their exclusive use.

Alvarez saw this as unimaginative. Although he was encouraged to claim some beam time as a reward for his labors, he came up with a better idea. Volunteering to track everyone else's efforts, he posted a cardboard wall chart of isotopes, driving a nail into every square where a radioisotope could exist. Whenever a new one was produced, he hung up a round tag that listed its properties. As a result, he learned it all by heart—a body of knowledge that would prove enormously valuable.

At the time of Alvarez's arrival, Lawrence had been gone on a

Alvarez's chart of isotopes at the Radiation Laboratory.
Lawrence Berkeley National Laboratory.

fundraising trip. On his return, he announced that he had secured money for a new cyclotron with a chamber diameter of sixty inches. It required a correspondingly massive magnet, which Lawrence impulsively ordered Alvarez to design. Seeing that it was a test, Alvarez admitted that he didn't know anything about magnets, to which Lawrence replied, "Don't worry, you'll learn."

Alvarez dutifully went to work at his bench, which was located in an undesirable spot near the mice. To determine the magnet's specifications, he had to build a scale model of each possible design, run it at the desired field, and plot the magnetization curve before it burned out. He turned innumerable pieces on a lathe, finding the balance between iron and copper that produced the most powerful field at the least cost, which guided the construction of the world's largest working magnet.

Having proved his usefulness, Alvarez felt more comfortable calling Lawrence by his first name. From his new mentor, he learned the value of intuition, which was often more helpful than technical knowledge, as well as the delicate art of fundraising. Lawrence accepted the importance of showmanship, and he delighted in stunts like bringing the beam "into the room." Turning off the lights, he invited visitors to look at the violet ray produced when charged particles struck the air outside the cyclotron, filling it with the smell of ozone.

On another occasion, Lawrence was showing off the cyclotron to John Cockcroft, an emissary from Lord Rutherford. He wanted to demonstrate what happened when the machine broke down, but for once, it refused to fail. Finally, he whispered to Alvarez, "I'll turn off the air and you give it the juice." To Alvarez's horror, Lawrence switched off the coolant system, causing the glass insulators to soften and collapse. Lawrence looked on with satisfaction as the crew rushed to the rescue, while Alvarez resigned himself to a long night of repairs.

Improving the cyclotron's maximum power was one of Lawrence's obsessions. He had found that the beam somehow became more intense when ovoid strips of iron, or shims, were inserted between the can and the magnet. Alvarez vividly remembered how Lawrence, in his orange lab coat, "would kneel for hours coaxing a little more beam out of the cyclotron," working by trial and error.

On the assumption that he would eventually be sent to build a cyclotron at another university, Alvarez diligently mastered the material. To understand the oscillator that produced the alternating current, he taught himself radio engineering from a textbook. Learning what was inside every drawer at the lab, he watched his more senior colleagues to figure out their tricks, reviewing it all in his head as he walked home to Euclid Avenue each night.

As a research fellow, he earned a decent $1,200 a year, but many researchers were men of independent means who gladly worked for free. There were no private rooms—everyone just had part of a workbench and a key to the front door—and the building was always full of activity. Scientists who put in only seventy hours a week were seen as "not very interested in physics," and most returned after dinner, while their wives, including Geraldine, were expected to sit there reading or knitting.

Alvarez saw the team approach to science as Lawrence's greatest invention, but he needed time to get used to it. One afternoon, he was trying to run the cyclotron when a crucial valve, which pumped cooling oil into the magnet coils, refused to work. He found that it had been closed by the interlock, a safety switch consisting of a weight on a length of clothesline. Instead of investigating further, he "buggered" the interlock with a clamp, forcing it open, and started the cyclotron as usual.

When he returned that night, the furious crew chief scolded him in front of Geraldine. The interlock, he learned, had deployed because a circulating pump had broken. By opening the valve, he had drained the oil, creating a sticky mess in the basement. Even worse, he had exposed the coils, which could have overheated, shutting the cyclotron down for weeks. He had been saved yet again by luck, and he was relieved when the chief didn't mention it to Lawrence.

Safety was a constant concern. Lawrence feared that the wooden building, which was soaked in transformer oil, would someday go up in flames. So much power was floating around that Alvarez could make a light bulb glow simply by touching it against a metal surface. A graduate student named Philip Abelson was once at the rear of the cyclotron, near the oscillator, when Lawrence turned it on without realizing that anyone was there. Alvarez built a switch to prevent a repeat occurrence, but the close call shook Abelson: "I was two or three inches away from frying."

A more insidious hazard was radiation. In the days before artificial radioactivity was discovered, the researchers in the lab had made Geiger counters, which suffered constantly from background noise. They assumed that they simply weren't good at building them, only to realize later that the counters were picking up the radiation that surrounded them at all times.

Lawrence's brother, John, had tried to make the others aware of the risks. In one test, he put a mouse in a brass cylinder next to the beryllium target in the cyclotron, which was producing neutrons under bombardment. After fifteen minutes, he showed them that the mouse was dead. Later, it emerged that it had actually died of suffocation, but the staff got the message. At John's urging, the control desk was moved to a safe distance and water tanks went up as shielding. In all likelihood, Alvarez later realized, these precautions saved his life.

Despite such measures, some degree of radiation exposure was taken for granted. Alvarez became friends with Willard Libby, a chemistry instructor, who often needed radioactive substances for research. In the past, Libby had produced the radioisotopes in his own lab, but he found that he could get better results just by leaving a sample in a window of the chemistry building, where it was thoroughly irradiated by the cyclotron next door—a full thirty feet away.

The culture in Berkeley encouraged the staff to take risks. Researchers routinely spent hours with their faces against the cyclotron window, troubleshooting the beam, and other exposures occurred by accident. One day, Alvarez wanted to show a visitor the beam inside the cyclotron, which happened to be using an older chamber that could run without a vacuum. When it contained air, as he thought it did now, the beam was visible as a ray of blue light.

Asking the crew to switch it on, Alvarez knelt with his friend by the window. When he didn't see anything, he yelled that there wasn't any beam, only to be informed that it should be there. Looking to one side, he saw the beryllium target glowing red. Alarmed, he yanked his guest away. He realized that someone had evacuated the chamber, making the beam invisible, and that they had been briefly exposed to the highest dose of fast neutrons in history.

On another occasion, Lawrence wanted a radiosodium sample for a lecture, so a target of sodium metal was bombarded around the clock. It was cooled by liquid air, which Alvarez was careful to add only when the cyclotron was off. Seeing another researcher blithely pouring a flask onto the target while the machine was running, Alvarez found that his colleague assumed, incorrectly, that no neutrons were being produced. Alvarez wrote with amazement, "This was undoubtedly the largest neutron dose that anyone had received to his genitals."

By the end of his first year, Alvarez was becoming "a jack of all trades" on the technical side, but he was mindful of the gaps in his knowledge, which he blamed on his inadequate education in Chicago.

"From another point of view, though, my training had been extraordinarily good. I could build anything out of metal or glass, and I had the enormous self-confidence to be expected of a Robinson Crusoe who had spent three years on a desert island."

It made him uniquely capable of surviving in the unstructured environment of Berkeley, where he finally felt ready for a real experiment. Lawrence was unmatched at dealing with engineering challenges, but he relied on others to come up with projects, which made him eager to have scientists who could find interesting problems on their own. To establish his reputation, Alvarez was determined to start with something impressive, and he would make sure that Lawrence noticed it.

AS ALVAREZ OBTAINED the theoretical knowledge that he needed to stand out from his peers, he found that you acquired good taste in physics in much the same way as you did in any other field—from reading, experience, and the right role models. What he described as his "coming of age as a nuclear physicist" was enabled by two indispensable resources, one widely utilized, the other comparatively neglected, and both crucial to his development.

The first was the physics community in Berkeley, which offered a refreshing contrast to Chicago. Of his university experience, Alvarez said, "Most of the graduate students didn't understand any quantum mechanics, largely for the reason that the professors had just learned it themselves and were so bewildered by the mathematics." Lawrence needed his researchers to be familiar with the latest developments, so he took steps to keep them from feeling isolated from the theorists next door.

His most significant innovation was a weekly journal club to discuss the latest papers. Wives avoided scheduling anything for Monday night, when all the scientists headed for the LeConte library. Staffers occupied chairs at the periphery, while senior physicists sat at tables at the center. The speaker—never announced in advance—could be anyone from a lowly graduate student to Robert Oppenheimer, who was building Berkeley into a powerhouse for theoretical physics.

Alvarez's other great resource was the physics library itself. Remembering how much he had learned from reading the primary sources, he undertook a systematic survey of the literature. Starting at 1919, the publication date of Rutherford's transmutation of nitrogen, he worked his way through five leading journals. Every night, after spending ten hours at the lab, he took home several years of issues, which didn't leave much time for Geraldine.

Scanning the tables of contents, he reviewed all the articles on nuclear physics, along with abstracts in German and French. The physicist Emilio Segrè, who arrived the following year, later pointed out that you could find the most important scientific papers by examining the edge of a volume of journals, looking for where it was stained by oils from the hands of many readers. Alvarez began to see why major experiments were done by certain people or labs but not others, and he kept an eye out for problems that would benefit from his tools.

His most valuable source was a set of three papers in *Reviews of Modern Physics* by Hans Bethe, Robert Bacher, and Stan Livingston. Bethe, a German immigrant who was seen as the country's leading theorist, felt that he understood a subject only after he wrote about it, so he summarized all of nuclear physics in what became known as "Bethe's bible." Alvarez, who paid for copies out of his own pocket, felt as though it had been written just for him. He carefully studied all five hundred pages, focusing on experiments that Bethe thought couldn't be done.

Early in 1937, Alvarez finally reserved some beam time at the cyclotron. Looking for an unexplored area, he tried to use alpha particles to turn lithium into boron, but accidentally made potassium instead. In hindsight, he felt that he had violated his own high standards. Producing yet another unremarkable isotope wasn't the kind of accomplishment that would get Lawrence's attention.

Alvarez was irritated by theorists who took credit for an idea without doing the hard work of the actual experiment, so he settled on another project—inspired by Bethe—to prove his point. It centered on beta decay, in which an unstable isotope gave up energy to produce a stable "daughter" product. In its most familiar form, an uncharged neutron in the nucleus decayed into a positive proton, emitting a negative electron and the supposedly massless particle called the neutrino.

Another kind of beta decay went in the other direction, with a proton decaying into a neutron and emitting a neutrino and a positron, the positively charged antiparticle of an electron. Alternatively, the proton could turn into a neutron by capturing an electron from the atom itself. Bethe noted that electrons in the shell, or orbit, closest to the nucleus were naturally the most likely to be captured. The innermost orbit was called the K shell, so the process was known as K-capture.

Bethe expressed doubts that electron capture could be observed directly, since it emitted only an undetectable neutrino. In a paper by another researcher, Alvarez saw one possible way. When an electron was captured by the nucleus, a vacancy appeared in the innermost orbit, allowing one of the electrons in an outer shell to "drop down" into the gap. When this took place, the atom emitted an X-ray at a wavelength equivalent to the difference in energy between the shells. Detecting it would prove that electron capture had taken place.

To find the elusive X-rays, which were usually lost in a sea of radiation, he had to set an elaborate trap. One advantage to the experiment—which had been tried unsuccessfully at the lab before—was that he could work it out at his leisure without access to the cyclotron. He was also unlikely to get extended beam time until he pulled off one big project on his own.

The "radioactivity factory" in Berkeley gave him access to a wide range of isotopes that were rarely put to use. As he considered how to detect electron capture, which was more common in heavy atoms, he reviewed the isotope chart, as if shopping from a catalog. He saw that positrons were emitted in beta decay after titanium was transmuted into vanadium, which came immediately after it on the periodic table. Alvarez thought that it was the perfect place to see electron capture as well.

Because the X-rays were so faint, he carefully planned every step. To bend the distracting positrons out of the way, he used a magnet that he had built to calculate the specifications for the new cyclotron. The radiation that remained was directed along a tube of helium, protecting it from absorption by the air, to a counter with thin cellophane walls, instead of the usual glass. It was filled with argon, which reacted only with X-rays, allowing the more energetic gamma rays to escape.

To confirm that the X-rays were being produced by electron capture in vanadium, he made a wooden frame to hold sheets of foil between the radiation source and the counter. By adding more sheets, he could plot the observed decline in the count against the curve that the element was expected to display. They matched precisely, convincing him that he had successfully observed electron capture.

Later, however, he realized that a similar result could be caused by a different process, called internal conversion. To rule this out, he invented a new test using radioactive gallium, which he obtained with the help of a chemistry postdoc named Glenn Seaborg. While waiting for it to decay in the basement of the physics department, he also befriended the Chinese graduate student Chien-Shiung "Gigi" Wu. "In addition to being an extraordinarily talented experimental physicist," he recalled, "she was one of the most beautiful girls I have ever known."

Lawrence—who needed researchers who made clever use of ordinary materials—was impressed by the results. As Alvarez was wrapping up, Lawrence suggested that he look for antiprotons, the particles that the theorist Paul Dirac predicted would have the mass of protons, but a negative charge. If they were produced at the cyclotron, they might show up in a cloud chamber, a detector filled with vapor that was ionized by charged particles in a visible trail.

Taking the hint, Alvarez built a narrow cloud chamber that fit unobtrusively between the cyclotron's poles. To find a particle's sign of charge, he deflected it with a magnetic field that he "borrowed" from the cyclotron, although the iron in his equipment weakened the beam. Since it couldn't be switched off, he and his colleagues were exposed to radiation the entire time: "We sat there with our heads close to the target, looking right into the cloud chamber waiting for the thing to pop."

In the end, he didn't see any antiprotons, which were unlikely to appear at the available energy levels, but he understood why Lawrence wanted to check. Having missed important discoveries before, he was determined not to make that mistake again: "Ernest believed that the way to find out about nature was to do experiments and see what turned up. Further, he made this suggestion when there was no credible theory to guide one's thinking."

Alvarez appreciated the need to look past the obvious, which was reinforced by some advice from his father. On one of his frequent visits to Berkeley, Walter Alvarez shared a cautionary tale. Decades earlier, he had met a woman who suffered from pernicious anemia, a lack of blood cells that was usually fatal. After failing to respond to conventional treatment, she had been cured by "her old Chinese nurse," who gave her a soup made with liver. Despite this clue, Walter failed to follow up, neglecting a line of research that later won one of his colleagues the Nobel Prize.

Walter blamed it on his tendency to keep his mind on the projects at hand, rather than allowing it to wander over "the full range of his work." He had once seen Will Mayo, the clinic's founder, seated quietly on his yacht on the Mississippi River. Mayo explained, "I think I am behaving wisely—sitting here, just thinking, because I am figuring out how I can strengthen certain weak sections of the clinic, and wondering how I can improve our public relations with the medical profession. I think I am of more use here, thinking and planning, than I would be at Rochester, operating."

Alvarez's father told him to follow Mayo's example. "He advised me to sit every few months in my reading chair for an entire evening, close my eyes, and try to think of new problems to solve." Like Walter, he was often too busy to take this advice to heart, but he finally put it into practice in the spring of 1937. A pair of gall bladder attacks, accompanied by jaundice, sidelined him for a week, and he benefited from the involuntary break. One evening he wrote in his journal, "I'm sitting down in my chair and I'm going to see if I can think up something useful tonight."

As he settled into his first meditation session, one leg slung over the arm of his chair, his mind turned to a series of talks that he had recently attended by the Danish physicist Niels Bohr. At first, Bohr had lectured to a packed house, but he was barely audible, lowering his voice even further to emphasize an important point. His audience quickly diminished, but Alvarez, who sat up front, was struck by Bohr's discussion of how a passing neutron could be picked up by an atom. In general, its likelihood of capture was inversely related to its energy, which made sense—the slower the neutron, the more time it spent in the atom's vicinity.

For different elements, the odds of picking up a neutron were highest at a specific velocity, which occurred at a threshold point called the neutron capture resonance. While it was possible to rank elements by this property in a relative way, no one had measured the absolute energy levels involved. Alvarez saw that he could derive these numbers from how long the neutrons took to travel a given distance, or their time of flight. Since neutrons of different speeds were jumbled up, however, he needed a counter that would ignore any neutrons that he didn't want to detect.

It was his first project that demanded extended use of the cyclotron. He modified the oscillator so that it switched rapidly on and off, producing a neutron pulse from the target sixty times per second. Beyond it, a counter would briefly turn on a few milliseconds after every burst. By then, the fastest neutrons would have passed, while the slowest ones would arrive after the detector shut down, so only neutrons in the desired range would be counted. Finding that he couldn't modulate the beam quickly enough for the neutrons that had originally drawn his attention, he pivoted to slower "thermal" neutrons, with energies equivalent to room temperature.

He was backing away from the problem of neutron capture resonance, but the results were promising for other reasons. To study nuclear reactions, experimenters needed neutron beams of a uniform energy. In his setup, the other energies were there, but because they weren't detected, he could act as though they didn't exist. It wasn't quite the pure beam that others were pursuing—it was actually a velocity selector, which picked up one energy level and tuned out everything else—but he sensed that it must be good for something.

In the summer of 1937, he attended a lecture series at Stanford by Enrico Fermi on slow neutron diffusion. "My friends and I attributed his flawless performances to careful rehearsal. Later, when I worked with him, I understood that he simply knew his subject so well that he could improvise a finished presentation only minutes before he walked into the room." Alvarez later joined the Italian physicist on Lawrence's cabin cruiser for a sightseeing tour of the bay. The following year, Fermi won the Nobel Prize for his work with slow neutrons, underlining the potential usefulness of Alvarez's beam, although he was still looking to prove its value.

Ultimately, it took him two more years to use it for "a quickie experiment." The physicist Edward Teller had shown that a molecule's interactions with passing particles were affected by the spin orientation—a form of angular momentum—of its protons. Alvarez decided to see how neutrons were scattered by two different types of hydrogen. In parahydrogen molecules, the protons had antiparallel spins, which canceled each other out; in orthohydrogen, they were parallel, reinforcing each other, so they affected nearby particles more strongly.

To measure the difference, Alvarez ran slow neutrons through ordinary hydrogen—which contained both kinds—and then through parahydrogen alone. Because the method was most effective at low temperatures, he obtained cold gas by boiling liquid hydrogen. He later learned to treat it with respect, since it could explode if mishandled, but for now, his team took a more casual approach. One day, when they had to dispose of an unused flask, Kenneth Pitzer, an associate professor of chemistry, simply poured it on the pavement, watching as it evaporated and drifted away.

At that point, Alvarez had long since proved himself as a creative experimenter. "By combining Bethe's theoretical insights with the experimental imagination that has been my distinguishing quality as a physicist, I suddenly surged forward from the back of the pack of young researchers." He was using the cyclotron in ways that surprised even Lawrence, who said that the "exciting physics" of the slow neutron beam was the most interesting work at the lab.

To confirm his newfound status, Lawrence invited Alvarez to sit at the center table at the weekly journal club. He was joined by Edwin McMillan, a slightly older colleague who had arrived two years before him. At Caltech, McMillan had mastered chemistry, theoretical physics, and electrical engineering, which made Alvarez feel insecure at his relative inexperience. Lawrence came to see them both as "his scientific sons and heirs," but McMillan had a head start.

Over the next few decades, Alvarez and McMillan would advance together at Berkeley, engaging in an unspoken rivalry. For now, they were widely viewed as the lab's most promising young researchers. Both had a knack for combining theoretical knowledge with workable experiments—the definition of good taste in physics. In achieving it, Alvarez had earned Lawrence's support, and he saw that the hardest part would be deciding what to do next.

AS ALVAREZ REFINED his slow neutron beam in 1937, he remained on the lookout for the "trouble" that Barton Hoag taught him to keep in mind. To make sure that the deuteron beam was really turning on and off sixty times per second, he ran the cyclotron without a target sample, exposing the detector to the deuterons alone. While he was watching the oscilloscope screen for pulses, he was surprised to see an unexplained burst of ionizing particles.

Eventually, he remembered that an electrode in the detector was smeared with lithium fluoride. When exposed to deuterons, the lithium was transmuted into an unstable isotope, producing a flurry of radiation. In other words, he had seen artificial radioactivity—the phenomenon that had rocked the scientific world in 1934. "I had just made a really extraordinary discovery," Alvarez recalled. "If I hadn't been three years too late, I would certainly have told someone about it."

This led him to wonder why it hadn't been discovered at Berkeley, where they had been using exactly the same equipment. At some point, it must have been seen—and ignored. He resolved not to make the same mistake: "I had now learned my trade and noticed things I hadn't expected to see, which is, of course, essential to scientific discovery."

To see the unexpected, he had to sustain the deliberate receptivity that his father advised, all for the sake of recognizing events that, by definition, were very rare. As it happened, the next one wouldn't occur for nearly two years. In the meantime, he was occupied by a project that depended, instead, on meticulous preparation. Once again, it was inspired by a theorist, who in this case literally came calling.

In the late spring of 1938, Alvarez was in the machine shop when the phone rang. Leaving his lathe to answer it, he found himself talking to Felix Bloch, a Stanford professor who wanted to measure the neutron's magnetic moment, or the strength of the magnetism produced by its spin. For charged particles, it could be calculated from their deflection in a magnetic field, but it was much harder with neutrons, which couldn't be uniformly accelerated.

Bloch eventually arrived at a method that required a reliable slow neutron beam, like Alvarez's velocity selector. Embracing it as a showpiece project, Alvarez was equally impressed by his new partner. Bloch spoke multiple languages, traveled widely, and played classical pieces on the piano before dinner at the faculty club. In one conversation, Bloch maintained that if you asked a group of physicists to rank the top hundred names in the field, they would come up with nearly identical lists. Real scientists, he said, were honest about their place in the pecking order.

In practice, Alvarez and Bloch each saw himself as the smartest man in the room, and the task ahead would test all of their combined skills. It called for polarizing neutrons—or lining up their spins—by passing them through magnetized iron, which acted like a pane of polarized glass. Next would come an oscillating field that depolarized some of them again, and finally a second pane that allowed only polarized neutrons through. By adjusting the settings to minimize the count on the far end, they could find the frequency that depolarized the most neutrons. They could then derive the magnetic moment with a simple equation—they just had to plug in the numbers.

It sounded straightforward, but the physicist Edward Purcell later called it one of the hardest experiments that anyone ever did. Alvarez concurred: "It certainly wasn't easy." On a limited budget, they acquired two enormous magnets, borrowing one from Shell Oil in Emeryville, where they loaded it onto a truck with a block and tackle. They learned to slow down neutrons with paraffin, so they didn't need the velocity selector after all, but a careless mistake by Alvarez cost them a month of work, which he recalled as "a pretty grim time." In the end, they counted two hundred million neutrons—more than had been previously detected in the entire world—over a grueling year.

When they prepared to publish, a clash of wills ensued. "We each thought our name should come first," Alvarez wrote. "It was Felix's idea—no doubt about that. But I made the points that A came before B, and additionally I had done *all* the work." The stakes were higher than a single paper. Alvarez wanted to underline the importance of conducting the experiment itself, rather than just coming up with the concept, and he eventually won the fight.

Decades afterward, Alvarez subtly put his colleague in his place, writing that Bloch—who was several years older—had been his "best pupil." Alvarez felt that he had taught Bloch how to be an experimentalist, but they never worked together again. Instead, Bloch followed up separately with groundbreaking research that laid the foundation for the field of magnetic resonance imagery. In 1952, Bloch jointly won the Nobel Prize with Purcell, beating Alvarez to the highest honor in physics.

Another achievement by Alvarez, which occurred during a pause in their work, was the result of pure luck. In 1939, the huge sixty-inch cyclotron was finally installed in the new Crocker Laboratory, but it initially ran at low power. To calculate how much shielding it required, Alvarez decided that it was easiest to turn it on, monitor its radioactivity, and install water tanks as needed. In the meantime, he wanted to find ways to use it at a modest level.

He came across a possibility on the same page of Bethe's bible that discussed electron capture. Lord Rutherford's team had bombarded deuterium—"heavy" hydrogen with one proton and one neutron—with deuterons, creating a fusion reaction that yielded ions with an atomic weight of three. These necessarily consisted of some combination of tritium, a hydrogen isotope with one proton and two neutrons, and helium-3, which had two protons and one neutron. One of them was radioactive, but they were moving too fast to tell which one it was.

For various reasons, most physicists thought that tritium was the stable one, while any helium-3 found in nature would have decayed long ago. No one had measured its half-life, or the time that it took for fifty percent of its atoms to decay, and in one of his meditation sessions, Alvarez came up with a simple but elegant method. He could make plenty of both isotopes in the old cyclotron—recently expanded to thirty-seven inches—by bombarding deuterium. Bringing the sample next door, he could set the new cyclotron to accelerate only helium-3. He could then take periodic radiation measurements over several years, calculating the half-life as the count fell.

Before he began, he needed to rule out a possible source of error. Alvarez had noticed that the cyclotron produced "junk ions" at certain frequencies, and if they occurred in the helium-3 region, the noise would ruin his measurements. He mentioned the problem to a graduate student, Robert Cornog, who agreed to help. After they wheeled the counter and oscilloscope over to the Crocker Laboratory, Cornog left to throw the hammer at an athletic meet.

Alvarez stayed behind to run a test of the sixty-inch cyclotron. Filling the chamber with ionized helium gas, the usual source of alpha particles, he told the engineers to decrease the magnetic field to three-quarters of its normal frequency, which was the level that he planned to use to accelerate helium-3. Watching the oscilloscope screen for junk ions, he didn't see anything out of the ordinary, which meant that the experiment would probably work.

"What followed that afternoon was one of the finest moments of my scientific life," Alvarez remembered. For one last run, he asked the crew to turn the magnetic field up to the standard level, and then he shouted, "Cut!" At this signal, the field was normally shut down only after the oscillator was switched off. This time, for some reason, the engineer cut the magnet while the oscillator was running, so it continued to accelerate particles as the magnetic field fell to zero.

Alvarez happened to be looking at the oscilloscope at just the right moment. To his surprise, the screen filled with a burst of pulses that quickly faded, followed by another. Running to the control room, he told the operators to do the same thing again, which produced an identical effect. He decided that the second burst was caused when the magnetic field fell into the region that accelerated protons, which were present as hydrogen impurities in the helium gas.

What mystified him was the earlier burst. It occurred partway between the frequencies that accelerated alpha particles and protons, implying that there was something in the gas canister with an atomic mass of three. Whatever was there showed up only when the magnetic field was changing, focusing it in the right region for a fraction of a second. On a hunch, he ran to get a set of calibrated aluminum foils, which he placed in front of the counter, making it just thick enough to block what he suspected—with growing excitement—was there.

When he checked the numbers, he saw that he was right. The cyclotron contained helium-3—the isotope that most physicists believed would never be found in nature. It had been there all along in the helium, which until recently had lain untouched for millions of years in natural gas fields in Texas. In other words, it had to be stable, or the opposite of what most researchers had assumed. By separating it out with the cyclotron, he had used the machine as a sensitive mass spectrometer, a novel application that would soon yield enormous dividends.

An even more noteworthy fact was staring him in the face. Alvarez was embarrassed that it didn't occur to him for over a week that if helium-3 was stable, then tritium obviously had to be radioactive. Long afterward, this isotope turned out to be tremendously important. During the Manhattan Project, the theorist Emil Konopinski saw that if tritium was mixed with deuterium, it could fuel a fusion reaction—the basis for the thermonuclear bomb. Alvarez knew that someone would have figured out tritium's properties soon enough, even without his help, but he took credit for his role: "I helped open that Pandora's box."

For now, Alvarez was delighted by his accidental breakthrough, which required less than a week of lab time. As he later noted, "A physicist gets his greatest satisfaction from finding something that is both unexpected and significant." It was a direct result of the freedom that he enjoyed at Berkeley. Now that the sixty-inch cyclotron was operational, he treated the older machine as his personal property, and he became annoyed whenever he had to stop what he was doing for outsiders.

These disruptions were usually due to medical research. Lawrence saw the use of radioisotopes in medicine as crucial to his mission, particularly for building goodwill with donors. Along with Lawrence's brother, John, the program was led by Dr. Joseph Hamilton, who became friends with Alvarez. At one point, Hamilton wanted to see how quickly radioactive sodium would be absorbed in the bloodstream, and he knew that the easiest way to find out was to swallow some.

Alvarez checked the calculations that night, confirming that it wouldn't be any higher than his radiation exposure on an average day. Bombarding table salt in the target chamber, they dissolved it in water and drank a hundred milliliters each. When it was Alvarez's turn, he downed the mixture and put his hand next to a Geiger counter in a lead box, which began clicking rapidly.

John Lawrence was more involved in neutron therapy for subjects with terminal cancer. For their visits, which halted all other experiments, he put on a neatly pressed doctor's coat. To avoid walking patients through the cluttered machine shop, a window was enlarged to create a separate door to the cyclotron. Ushered down a white plywood corridor to a temporary cubicle, they were restrained with their heads near the beryllium target.

The treatment turned out to be ineffective, but Ernest Lawrence continued to expand the cyclotron to make it more useful for medicine, which often required pausing it for upgrades. Convinced that this was hindering research, Alvarez prepared a chart of how the number of papers produced at the lab had fallen. Lawrence pointed out that Alvarez himself had published plenty of results during that period, implying that an inventive scientist would learn to deal with it.

Alvarez conceded the point, knowing that it was better to defer now to get his own way later. By then, his ambition and impatience were obvious to everyone. On arriving in Berkeley, Emilio Segrè received a candid briefing from Lorenzo Emo Capodilista, an Italian count who was one of the lab's gentleman scientists. He described McMillan as "very clever, but lazy," while Alvarez was "a little fascist leader, fawning to the Duce, but mean to his equals or inferiors."

His growing profile led to occasional job offers, including one from Chicago. To discourage him from leaving, Alvarez was promoted from instructor to assistant professor. He agreed to remain—perhaps to the disappointment of Geraldine, who missed her parents—and assumed a more prominent role. In a glowing recommendation, Raymond T. Birge, the department chair, wrote:

> [In 1937, Alvarez] seemed to have been about the most sought after young physicist in the country. Since then our expectations have been repaid with compound interest. We have never had in this department a young man so extraordinarily fertile in new ideas of real value. They come so fast that it is quite impossible to provide time or materials to carry them out. . . . [The slow neutron beam] is probably the most important instrumental advance in the field of nuclear physics since the invention of the cyclotron, and it will make Dr. Alvarez internationally known.

At twenty-seven, his future seemed clear. As it turned out, it would be disrupted by unforeseen developments, both on a global scale and in the esoteric world of physics, which were about to dramatically converge.

On the morning of January 30, 1939, Alvarez was getting his hair cut in the campus barbershop, leafing through the *San Francisco Chronicle*, when he came across a wire story on an inside page. It described an announcement from a recent conference held in Washington, DC.

A German chemist named Otto Hahn had bombarded uranium with neutrons to make barium, which had roughly half of its parent element's atomic weight. This made sense only if Hahn had split the atom, a reaction previously thought to be impossible.

According to the article, the newly discovered phenomenon could generate tremendous energy, but it had an even more immediate significance to Alvarez. Telling the barber to stop, he rushed outside, his hair only partially cut, and ran to the lab. The first person he saw was Phil Abelson, one of his graduate students, who had been working along the same lines. Alvarez was afraid that the news would give Abelson a heart attack. "I have something terribly important to tell you. I think you should lie down on the table."

Abelson complied, realizing numbly as he listened that he had missed making history by just a few days. As the other scientists showed up, Alvarez told them what he had read. Incredibly, it was the first that anyone in Berkeley had heard of a discovery that had been the talk of physics circles in Princeton and New York for two full weeks. At one point, Alvarez headed next door to LeConte Hall, looking until he found the one person in the physics department who needed to know about it at once. It was J. Robert Oppenheimer.

3.

UNDER THE HOOD

1939–1943

ONE EVENING IN 1934, ROBERT OPPENHEIMER DROVE TO the Berkeley hills with his girlfriend, Melba Phillips, who was also one of his graduate students. After returning from a solitary stroll, he leaned into the car window to tell Phillips that he would walk home on his own. Unfortunately, he didn't notice that she had fallen asleep. When she awoke, alone, she flagged down a policeman, who initiated a search for Oppenheimer—or perhaps his body. He was finally found in bed at the faculty club, where he apologized for having forgotten about Phillips: "I'm awfully erratic, you know."

After the incident made the papers, the article was posted on a bulletin board at the University of Chicago, which was where Alvarez first saw Oppenheimer's name. They met in Berkeley in 1936, when "Oppie" was thirty-two. With his unruly hair, compulsive smoking, and angular features—like the early illustrations of Pinocchio—he was an easy target for caricature. At a departmental picnic, graduate student Philip Morrison performed a comedy routine, lighting cigarettes while wearing a mop on his head, and everyone recognized the impersonation at once.

At the time, Oppenheimer was a Berkeley professor with a special dispensation to spend six weeks every year at Caltech. Alvarez credited him with figuring out how to explain quantum physics to students, but he wasn't a very good lecturer—he spoke quietly, went fast, and could be sarcastic and cruel, although he was also capable of memorable flourishes. At Lawrence's request, he performed a version of Alvarez and Hamilton's radiosodium test for an audience, swallowing a radioactive solution and waving his hand to set off a Geiger counter.

UC Berkeley physics department and Radiation Laboratory staff at the sixty-inch cyclotron magnet in 1938. Back row, left to right: Alex S. Langsdorf Jr., Sam Simmons, Joseph G. Hamilton, David H. Sloan, J. Robert Oppenheimer, William Brobeck, Robert Cornog, Robert R. Wilson, Eugene Viez, J. J. Livingood. Second row: John Backus, Wilfrid B. Mann, Paul C. Aebersold, Edwin McMillan, Ernest Lyman, Martin D. Kamen, D. C. Kalbfell, W. W. Salisbury. Front row: John H. Lawrence, Robert Serber, Franz N. D. Kurie, Raymond T. Birge, Ernest O. Lawrence, Donald Cooksey, Arthur H. Snell, Alvarez, Philip H. Abelson. National Archives.

Alvarez praised the graduate students under Oppenheimer—who he later said would have won the Nobel Prize if he had lived to see confirmation of his work in astrophysics—as "probably the finest group of young theorists in the world." In a smoky room in LeConte, they could be seen tossing out comments as one scribbled equations on a chalkboard. "Robert's permanently floating crap game was in every sense the theoretical counterpart of the Radiation Laboratory; the theorists similarly worked together, helping each other enthusiastically and happily."

Lawrence—whose second son, Robert, was named after Oppenheimer—was equally taken by the man who shared the center table with him at the Monday evening seminars. In contrast to Lawrence's humble background, Oppenheimer was a cosmopolitan figure who had traveled widely, studied philosophy, and learned foreign languages, although, Alvarez noted, "not much of the Sanskrit for which he is famous." Their only real points of disagreement were over politics. One day, Alvarez was at the controls of the cyclotron when Oppenheimer wrote an invitation on the blackboard to a benefit for leftists in the Spanish Civil War. On seeing it, Lawrence erased it without a word.

Oppenheimer was a mediocre experimentalist who couldn't solder two wires together, but Alvarez—like Lawrence—took full advantage of his theoretical brilliance, thanking him in his paper on electron capture. One night, while walking to the journal club, Alvarez suddenly pictured a new kind of cyclotron, which would route particles in a spiral across a single accelerating gap, instead of two. After he explained the idea at the library, Oppenheimer worked out the math on the spot for what became known as the microtron, although it wouldn't be built until after the war.

At his house, Oppenheimer occasionally hosted Alvarez and Geraldine for dinner, and he even took them once to the theater, where they were dismayed to find themselves watching a play about the labor movement. Alvarez still didn't know much about Oppie's background, let alone his complicated personal life. When Lawrence casually mentioned one day that Oppenheimer was Jewish, Alvarez expressed surprise: "I'd never thought about it."

On January 30, 1939, when Alvarez told the "bullpen" of theorists in LeConte about the newspaper article on nuclear fission, Oppenheimer replied, "That's impossible." At the chalkboard, he quickly demonstrated that the energy barriers in the nucleus were too high for the reaction to occur. Alvarez left without persuading him, but he didn't blame Oppenheimer for being skeptical. He was furious at himself for having missed the discovery, too, and his only consolation was that everyone else had made the same mistake.

Even Fermi, the brightest of them all, had failed to see it. After bombarding uranium with neutrons, he had concluded that the products were transuranic elements, which were higher up on the periodic table. Whenever Alvarez looked at his isotope chart, however, he felt that something "fishy" was happening. Transmutations usually occurred "downhill," along what was known as the valley of stability, and Fermi's reactions were going in the wrong direction.

Alvarez had spent several evening meditation sessions unsuccessfully searching for the answer, while Lawrence encouraged Abelson to look into the problem as well. When the news about fission came out of Washington, Abelson was on the verge of a solution, using a method that Alvarez had proposed. Armed with the latest details, he switched to a simpler technique—also developed by Alvarez—that confirmed the theory at once. As Alvarez later noted, "[Abelson] was so close to discovering fission that it was almost pitiful."

That evening, the Monday journal club was dominated by a heated discussion of fission, with Alvarez at the center table, still with just half of a haircut. Glenn Seaborg wandered the streets afterward for hours, fuming that he had overlooked something so obvious, while Oppenheimer said that he would be convinced only by confirmation from multiple labs.

Alvarez was eager to provide it. The next day, he cabled George Gamow, the George Washington University professor who had organized the conference where the announcement was made, and received some valuable advice. When bombarding uranium, Gamow said, the physicist Otto Frisch had seen large bursts of radioactivity, which clearly indicated that fission was taking place.

It was all that Alvarez needed. He placed a uranium sample in an ionization chamber, which would detect any radiation, exposed it to neutrons, and hooked it up to an oscilloscope. On the screen, he immediately saw a background of alpha particles from natural decay, along with enormous vertical pulses that could be caused only by nuclear fragments.

Looking at the scope, Alvarez realized that he had been even closer than Abelson to discovering it. The year before, he had used a nearly identical setup to search for long-range alpha particles from bombarded uranium. To eliminate the distracting background of short-range alphas that would also be there, he covered the uranium with foil, blocking them from the detector—and also shielding the counter from the bursts caused by fission, which he never saw.

Alvarez was convinced that if he had only changed the foil at the detector that day, he would have seen the fission pulses, which probably would have won him the Nobel Prize the following year. In retrospect, he felt that he was lucky to have missed it: "I doubt if I had the maturity at twenty-seven to handle the burden of having made one of modern science's greatest discoveries."

It was useless to brood over it now. Calling in Oppenheimer, he pointed to the pulses. "In less than fifteen minutes," Alvarez wrote in his memoirs, "he not only agreed that the reaction was authentic but also speculated that in the process extra neutrons would boil off that could be used to split more uranium atoms and thereby generate power or make bombs. It was amazing to see how rapidly his mind worked, and he came to the right conclusions." In an earlier draft, however, he wasn't sure if they discussed the bomb: "I don't remember the conversation in that much detail."

Oppenheimer undoubtedly conceded that he had been wrong, which Alvarez admired as "the scientific ethic at its best." On a more practical level, Oppenheimer—who had been mulling it over for a full day—had no desire to fall behind as attention turned to the possibility of a chain reaction. If two or more neutrons were released by fission, the collisions could spread in a geometric progression. What they lacked so far was proof of the crucial secondary neutrons, which would be lost in the neutron cloud that caused fission in the first place.

One day, Alvarez saw a simple way to find them. Going to the chemistry storeroom, he signed out some bottles of uranium oxide. At the lab, he tuned his slow neutron beam to yield only thermal neutrons from the cyclotron. He then surrounded the counter with paraffin and cadmium, blocking the slow neutrons as well, so the count fell completely to zero.

"My reasoning was unimpeachable," Alvarez recalled. Placing the uranium between the neutron beam and the detector, he kept his eye on the counter. If the count rose even the slightest bit, it had to be from secondary neutrons produced by fission, since he had carefully screened out everything else. After failing to see anything for five minutes, he turned to a different experiment.

It was perhaps the worst mistake of his entire career. The apparatus, he realized afterward, had been almost too easy to set up. If it had been harder, he might have taken the trouble to adjust the equipment to make it a million times more sensitive—by reducing the distance, adding more uranium and paraffin, or running it for longer. "I would have seen the secondary neutrons the same day. But I didn't; I was stupid. I didn't understand how important the experiment was."

Alvarez regretted not mentioning the idea to Oppenheimer, who would have told him to keep going, and he was equally unsuccessful in a later attempt to measure the heat of irradiated uranium. He was having trouble navigating the shifting landscape, partly because it was hard to discard habits that had worked in the past. As he continued to focus on problems inspired by Bethe, for the first time in years, the most interesting work at the lab was occurring elsewhere.

One line was driven by McMillan, who discovered traces of an actual transuranic isotope in the products of fission. To transmute enough uranium to find the even more elusive element 94, which he and Alvarez thought might be fissionable, McMillan had to push the cyclotron to the breaking point. Describing the result as "the most heavily bombarded sample in history," Alvarez realized that they needed more efficient ways to induce transmutations.

To produce a more massive "bullet," he filled the thirty-seven-inch cyclotron with methane and tuned it to accelerate carbon ions. Finding only a weak beam, he lost interest. Within a few years, Lawrence would use a similar process—electromagnetic isotope separation—to tremendous effect for the Manhattan Project, while McMillan and Seaborg followed a different path to find element 94, or plutonium.

Despite his fondness for "quickie" experiments that he could throw together as soon as they occurred to him, Alvarez was occasionally willing to do more. In October 1939, he heard a talk at the journal club by the spectroscopist W. Ewart Williams, who noted that a red spectral line from a cadmium lamp was a standard unit of length. For more precise measurements, however, a shorter wave would be best. When Alvarez asked about its ideal specifications, Williams advised using a pure, heavy isotope that yielded a simple line.

There was no known means of obtaining it with chemical separation, but Alvarez saw another way. That night, he pulled out his isotope chart, which he used as a kind of cookbook. Gold, he realized, could be transmuted into mercury-198, which had all the right properties. With the help of a graduate student, Jacob Wiens, he left a gold cylinder in a tube near the cyclotron, exposing it to neutrons for a month. When Wiens separated out the mercury vapor, it produced a beautiful green line that became a standard of length for over a decade.

Alvarez preferred to hand such ideas off to others, but when he saw a potential upside, he took greater pains. Reading a paper by Fermi on the absorption of charged particles, he checked it experimentally on his own time. On a visit, Fermi—a brilliant theorist who had also turned himself into a superb experimentalist—asked to help, taking measurements as Alvarez recorded them in his notebook. It reminded him that he needed mentors like this, although his most important patron would come from an unexpected direction.

In 1939, Alfred Loomis arrived in Berkeley by limousine. Loomis, who was fifty-one, had studied law at Harvard and earned millions in investment banking. His first love was science—he taught himself enough ballistics to research artillery during the Great War—and he spent a fortune on a private hilltop lab near his mansion in Tuxedo Park, New York. Granting scientists lavish resources, he personally financed projects ranging from radar to brain waves.

When he came to help raise money for a larger cyclotron in the Berkeley hills, Alvarez showed him around campus. During office hours, he held long discussions of nuclear physics with Loomis, who was an exquisitely polished version of the man Alvarez wanted to become—a compulsive tinkerer, full of ideas, and quick to master any subject. Like Lawrence, Alvarez was fascinated by Loomis's wealth, sophistication, and patronage of young scientists who could expand his knowledge, describing him afterward as "the last of the great amateurs of science."

Party to celebrate Lawrence's Nobel Prize at DiBiasi's Restaurant in Albany, California, in November 1939. Back row, left to right: Robert Cornog, Ernest O. Lawrence, Luis W. Alvarez, Molly Lawrence, Emilio Segrè. Second row: Geraldine Alvarez, Betty Thornton, Paul Aebersold, Iva Dee Hiatt, Edwin McMillan, Bill Farley. First row: Donald Cooksey, Robert Thornton, Bob Sihlis. National Archives.

The following year, as war intensified in Europe, Loomis returned. Hoping that the lab could aid in national defense, he passed along an idea from the FBI to mark classified papers with a radioactive tracer, which would enable security officers to detect whether someone was stealing documents. At Lawrence's request, Alvarez and a colleague, Bill Baker, built a Geiger counter small enough to slip into a book or pocket. It was never used, but it made a favorable impression at just the right time.

In October 1940, Loomis invited Lawrence—who had won the Nobel Prize the year before—to a secret conference in Tuxedo Park. Loomis was a contact for a mission from the United Kingdom, led by the chemist Henry Tizard, that was authorized to share information on British defense programs. In a black metal deed box, Tizard brought plans relating to the nuclear bomb and the proximity fuze, as well as a prototype of a device crucial to the invention that would have the greatest impact of all on the war—microwave radar.

To track enemy planes, the British had built the Chain Home radar stations, a network of steel towers that played a pivotal role in the duel with Germany in the Battle of Britain. The next step was an accurate radar set small enough to put on an aircraft. It became possible with the cavity magnetron, a vacuum tube that generated powerful microwave pulses in a circular chamber. Overnight, airborne radar seemed within reach, but with most British scientists busy elsewhere, Prime Minister Winston Churchill approved an overture to the Americans.

At Loomis's mansion, Lawrence and Vannevar Bush—the head of the recently established National Defense Research Committee—studied the blueprints on the living room floor. Lawrence declined a chance to run the project at MIT, but he pledged to enlist his best physicists. His staff had overseen huge technical projects under pressure, and they were comfortable with pulsed signals, unlike radio engineers, who usually saw them only when something had gone wrong.

First on his list were Alvarez and McMillan. Back in Berkeley, Lawrence hinted that he had big news involving techniques that they knew well. "I correctly assumed he meant electronic pulses," Alvarez wrote, "but I couldn't think of any military application." After they were cleared to hear the rest, McMillan remembered, they were excited by Lawrence's declaration "that this great project was starting and that we must get into it—that Hitler must be stopped."

With Lawrence's encouragement, Alvarez began brainstorming designs for a sensitive microwave detector. He had recently moved to Panoramic Way in Berkeley, where Geraldine planted a vegetable garden and prepared for the arrival of their first child. Walter Smithwick Alvarez was born on October 3, 1940, less than a month before the offer to join the radar program.

Although it would take him away from his son, Alvarez agreed to go. At that point, he expected that the MIT lab would be supplemented by a rotating roster of scientists, like himself and McMillan, who would stay for just a few weeks. In the meantime, Geraldine's recently widowed mother would help with the baby, while Alvarez planned to be home by Christmas.

On the afternoon of November 11, 1940, Alvarez and McMillan boarded their train to the patriotic sounds of a brass band. Lawrence saw them off at the station, where Oppenheimer—newly married to the former Kitty Harrison—handed each of them a bottle of whiskey for the journey. He used to give books to travelers, he said, but had stopped after the physicist Paul Dirac turned down the gift, explaining that reading "interfered with thought."

As the train pulled away, Alvarez sensed that he was leaving his old life behind. By giving up his resources in Berkeley for the sake of an urgent problem, he could regain his professional bearings. In the long run, it saved him, but he still recalled his departure with a touch of wishful thinking: "I was in the midst of my most productive years as a working scientist. At the time I didn't think about the physics I might have done had I stayed at Berkeley. That loss can never be restored."

IN SUMMARIZING THE EVENTFUL YEARS that followed, Lee DuBridge, the physicist recruited by Lawrence to run the MIT lab, later said, "Radar won the war; the atomic bomb ended it." One of the program's scientists, I. I. Rabi, declined the associate directorship of Los Alamos in favor of the radar project, telling Oppenheimer, "I'm very serious about this war. We could lose it with insufficient radar."

When Alvarez and McMillan arrived in Cambridge, it seemed likely that the brutal air engagement with Germany would play a greater role in victory or defeat than the atomic bomb, but they had to move fast. On their first day at the MIT Radiation Lab—a name intended to deflect suspicion, since no one could be looking seriously into nuclear weapons—they found just half a dozen rooms on two floors of a nondescript building, with a single guard stationed at the door.

Alvarez's initial task was to educate himself. He knew a lot about radio engineering, but not antenna design, which hadn't been relevant to the cyclotron. To make a contribution now, he had to learn quickly, while still keeping up with physics—or so he assumed. In the end, however, he attended just one meeting of the campus journal club. Radar, which had attracted the best people he knew, demanded all his attention: "In a few weeks I felt sorry for my M.I.T. colleagues who were stuck in nuclear physics when they could have been working on radar."

On November 15, his first official meeting at the Rad Lab began with a mission overview from Edward George "Taffy" Bowen, the legendary engineer who led the airborne radar group in Britain. Their priority, Bowen said, was fighting at night. Because of their short range, existing radar sets on airplanes had to be supplemented by operators on the ground. Despite its limitations, the system was so effective that a rumor was spread—to put the Germans on the wrong track—that RAF pilots improved their night vision by eating carrots.

Installing precision radar on the planes themselves could be done only with microwaves. Alvarez was assigned to Rabi's group, where he built an enlarged cavity magnetron for testing, and he assumed a greater role after a visit from Loomis and Lawrence. Writing a timeline on the blackboard, Loomis, who was heading the NDRC microwave committee, called for a functional rooftop set by January.

To serve as head of production, Loomis chose Alvarez, who promptly recruited Lawrence Johnston, his teaching assistant in Berkeley, and became friends with DuBridge—the unassuming Oppenheimer of radar—while carpooling to work to save on gas. Realizing that he would be left behind if he went back to California, he arranged for Geraldine and Walt to join him in an apartment in Cambridge. As she shipped their furniture and said goodbye to her mother, Geraldine had no way of knowing that they wouldn't return to Berkeley for five years.

Using a sprawling setup on the roof, Alvarez's team obtained a radar reflection from the dome of the Christian Science church in Boston in January 1941. Scaling it down for a plane would be more challenging. Because everything had to fit in the nose compartment, the same antenna had to function as both a receiver and a transmitter. The former was often fried by the latter, and even after this issue was resolved, background noise made it hard to see anything at all.

On his next visit, Lawrence was unimpressed. After hearing that they had yet to detect any planes, he advised them to scrap the system and start over. Alvarez was upset—he felt that Lawrence didn't appreciate their hard work—and offered to bet that it would succeed. A meeting in Washington would decide its fate in two days, so they embarked on a frantic round of improvements.

Early on February 7, 1941, Alvarez returned to the roof with a skeleton crew. They suspected that their lingering issues were due to the elevation mechanism that controlled the antenna, so an engineer simply detached it, aiming it by hand, as another man looked through a telescopic sight for passing aircraft. In the tarpaper radar shack, Alvarez suddenly picked up a blip on the scope. Rushing out to check the sky, he saw a plane overhead. He ran to call DuBridge in Washington, who proudly announced to the committee, "We've done it, boys."

Lawrence sent a congratulatory telegram—"I had my words for lunch"—but they still had to modify the system for planes. Alvarez built a plywood mock-up of the compartment of a B-18 bomber to develop the prototype, which was followed by installation and testing in the air. On March 27, Alvarez, Bowen, and McMillan took off with the radar set, cheering as it picked up a target plane and saw ships from eight miles at Cape Cod. Flying on to New London, Connecticut, they detected navy submarines on the surface.

The strong performance at sea—unhindered by echoes from the water—was an unexpected bonus. As the Battle of Britain ended, night fighting became less of a priority, but U-boats remained a serious threat. Along with intercepts from the Enigma cipher machine, radar was a crucial weapon, leading to a war of countermeasures. After the Germans learned to detect beams from British planes, submarine crews were trained to dive before the fighter could open fire.

To outmaneuver the Germans, Alvarez eventually came up with a clever ruse. When an airplane found a submarine, it could decrease the power of its outgoing pulses, while still receiving a strong enough signal in return to find its target. The diminishing beam would trick the sub into thinking that the plane was moving away, allowing the fighter to draw within firing range. Because the system "foxed" the U-boats, it became known as Vixen.

Long afterward, Alvarez described the result with satisfaction: "I don't know when Vixen became operational. I was on the Manhattan Project at the time. But a number of people have told me that it had a devastating effect on U-boat operations. Things happened so quickly that the doomed sub wasn't able to tell headquarters what had happened. So all they knew was that lots of subs simply disappeared. That was certainly bad for morale."

IN MAY 1941, as Alvarez was preparing to tour radar labs in the United Kingdom, he was awakened one morning by chest pains. After an examination revealed gallstones, he arranged for an operation at the Mayo Clinic, driving there with Geraldine as Walt slept in a bassinet. Although the procedure was a success, two weeks in bed—he was nauseated by the anesthesia—led to phlebitis. The inflammation of his leg veins, which persisted for the rest of his life, was eased by medication and elastic stockings, but it became painful to stand at cocktail parties.

After recuperating over the summer—he spent it tinkering with a method for detecting gallstones with X-rays—he found that the Rad Lab had advanced tremendously in his absence. Just as his earlier health scare in Berkeley had forced him to slow down and think, he treated the involuntary withdrawal as an opportunity. With everyone racing to keep up with Germany, it was easy to forget that they possessed an incredible tool—microwave radar—that had yet to be fully explored. Now he had time to dream up applications that hadn't occurred to anyone else.

His most significant idea was inspired by the SCR-584, a ground radar system for fire control. As it automatically tracked the coordinates of an enemy plane, a computer calculated the trajectory for shells to hit the target. Watching a prototype on the roof in August, it struck him that instead of aiming guns, it could guide pilots in blind landings. Because of damage or low fuel, planes often had to land in bad weather or at night. Alvarez assumed that the system could be modified to bring them safely to the ground, but the actual process consumed much of the next two years.

In the meantime, he searched relentlessly for other uses for microwave radar. He began with an idea that he could develop himself. To tell friendly pilots from enemy aircraft, the British used a beacon system, with each plane carrying a transponder that responded to radar pulses with a signal in Morse code. It took several seconds to transmit, but Alvarez saw that a microwave beacon could reduce this time to microseconds, displaying the code as a pattern of light and dark bands—recognizable at a glance—on the recipient's radar scope.

This early success emboldened him to tackle larger projects. Avoiding Loomis's attempt to put him in charge of radar countermeasures, he preserved his involvement in design as head of the Jamming, Beacons, and Blind Landing group. Eventually, he managed to avoid any categorization at all. While other groups focused on specific components, he was authorized to develop his own ideas in Division 7, or Special Systems, nicknamed "Luie's Gadgets."

One project came out of a discussion with Bowen in November 1941. They had been debating the shortcomings of the Norden bombsight, the mechanical computer that calculated a bomb's point of impact. As an optical sight, it couldn't be used at night or in cloud cover, and it called for the bomber to maintain a fixed course, making it vulnerable to attack. Hoping to replace it with a radar system, Alvarez came up with a fanlike beam that could scan the ground from side to side.

To generate a beam that was narrow horizontally, but with a wide vertical angle, the antenna needed the opposite proportions, so it had to be long and thin. Alvarez proposed installing it on the wing's leading edge, which called for an unconventional design. Since it couldn't be uniformly powered by a single antenna at the center, he envisioned a line of tiny antennas along the wing itself.

According to the engineers he consulted, this was impossible—the antennas had to be separately tuned to prevent interference, which became unworkable with more than four. Thinking over the issue,

Alvarez saw an analogy to classical mechanics. Famously, there was no known way to use Newton's laws to predict the motion of a system of more than two bodies—the notorious three-body problem. If many particles were involved, however, it became a simple matter of thermodynamics.

In the diffraction gratings that he had studied in college, for example, the interference produced by thousands of rulings could be easily analyzed with statistical methods. Alvarez guessed that a similar principle would apply here: "I just predicted that if you go to very large numbers then there are no problems at all and it turned out to be true. Apparently this had not occurred to the radio engineers. But it occurred to someone who had worked with diffraction gratings."

His theoretical inexperience turned out to be an advantage as he designed the "leaky pipe" antenna, a metal waveguide with regular slots that radiated energy from an oscillating current. What became known as the Eagle bombsight generated sharp images, but it was delayed by technical issues until the end of World War II, when it was used in raids against Japanese oil refineries. A history of the lab concluded, "Eagle was a good set. It just ran out of war."

As usual, Alvarez was already looking for ways to exploit the same principle. In May 1942, he proposed installing an early warning system on the West Coast to watch for air raids from Japan. Until then, microwaves had been used to reduce the size of equipment, but they could also increase the range of larger antennas. Previous efforts had been based on parabolic dishes, which generated a narrow beam that was unsuited for searching the sky.

Alvarez proposed an antenna with an enormous reflector—shaped like a billboard and powered by a clever arrangement of alternately oriented antennas—that could produce a fanlike beam, mounted on a circular track to scan the entire horizon. While it never made it to the home front, a version of the Microwave Early Warning system monitored the airspace during the Normandy invasion and defended against V-1 bombs in England. Like the Eagle, the system once dismissed as "Alvarez's folly" proved its value, but only when the war was nearly over.

ALL THE WHILE, Alvarez was closely following the work on the atomic bomb. In the autumn of 1941, Oppenheimer had invited him to Columbia for a meeting on the production of fissionable isotopes. Although he wasn't officially cleared, Alvarez examined the uranium pile built by Fermi, which used graphite blocks as moderators to slow down neutrons enough to potentially start a chain reaction.

Lines of power were forming across the country, along with a growing debate over how to consolidate the project. At Columbia, Harold Urey oversaw the production of fissile uranium-235. In Berkeley, Lawrence was converting the cyclotron into an electromagnetic isotope separator, or calutron, drawing on the principles that Alvarez had used to turn it into a mass spectrometer. Finally, in Chicago, Arthur Compton's lab was preparing to build reactors to make fissionable plutonium.

Alvarez was pulled into the discussion a few months later, when Compton called to ask for his advice on the isotope program. In Chicago, he participated in a tense meeting attended by Lawrence and Leo Szilard, the Hungarian physicist who had been the single most fervent advocate for the bomb. As Compton lay in bed with the flu, Alvarez's two mentors each argued for their own university. After Compton declared that he was bringing Fermi to Chicago, Lawrence fired back. "The whole tempo of the University of Chicago is too slow."

Compton bristled. "We'll have the chain reaction going here by the end of the year."

"I'll bet you a thousand dollars you won't," Lawrence replied. Alvarez, who had been pushing for MIT, was startled by the exchange: "Arthur thought for a minute and announced that he would take the bet. Then they both looked at me, obviously embarrassed that I had heard these intemperate remarks." In the end, they reduced the wager to "a five-cent cigar," which Compton eventually won, after Fermi produced a controlled chain reaction in a squash court in Chicago.

Although Alvarez's opinions were respected, he still wasn't formally involved. When the atomic weapons lab was established under

Oppenheimer in Los Alamos, New Mexico, Hans Bethe proposed him as a recruit, only to be told by Edward Teller that "[they] were not very eager to have Alvarez." For now, his radar work seemed more important, and, as the historian Gregg Herken observed, "his oversized ego may also have been a factor against it."

On December 7, 1941, Alvarez was reading the paper in Belmont, where he and Geraldine were renting a comfortable house, when news came over the radio of the attack on Pearl Harbor. His first impulse was to reassure Geraldine that Japanese bombers could never have reached the base—the antennas there would have seen them from a hundred miles away. He was later horrified to learn that a massive formation had indeed been detected on radar, but the warning had been disregarded. Technology, he saw, was only as useful as the people who were responsible for it.

Ironically, Alvarez remembered the year that followed as one of the happiest times of his life, apart from a private tragedy. Geraldine became pregnant again—they planned to have children two years apart—and went into labor in the fall of 1942. She had previously complained that her obstetrician at Massachusetts General was a "cold fish," and while Alvarez was in the waiting room, the doctor showed up to tell him brusquely that the baby was dead. "Having delivered his message," Alvarez wrote, "he turned on his heel and walked away."

They were devastated by the loss of the baby, a girl whom they named Jean, and Geraldine was forced to stay in the care of the doctor who had shared the news so coldly. Alvarez recalled, "Gerry later extracted from him that, in the process of adaptive switching from placentally refreshed blood to blood that went through baby Jean's lungs, a critical valve did not function and the baby suffocated."

He had no choice but to return to the lab, leaving him with little time to comfort his grieving wife. If he stepped away now, he would lose all the momentum that he had built, and the situation was worsened by his inability to discuss his work with Geraldine—or anyone beyond his own circle. He learned to tell plausible lies, since a denial would only arouse suspicion, and figured out how to mislead knowledgeable outsiders, often based on a single piece of information that they lacked.

When his friends asked about airborne radar, for example, he said that it wasn't possible with the available wavelengths, taking advantage of their ignorance of the cavity magnetron: "Let's do the numbers." He found a similar way to shut down discussion of the atomic bomb. It was widely assumed that only slow neutrons could produce a chain reaction, ruling out the conditions required for an explosion. Alvarez knew that fission could be induced in U-235 by neutrons at any speed, but an obsolete argument could still convince intelligent experts that it didn't make sense.

It seemed like a necessary violation of the truth, for the sake of the greater good, and he reserved it for special occasions: "The only times I have consciously lied about substantive matters have involved national security." Refusing to comment wasn't enough. He needed to invent a viable alternative, which was rarely an issue for a man of his ingenuity. It was a skill that he would often use again, and he was always persuasive: "I never had a dissatisfied customer."

WHEN ALVAREZ CONCEIVED of a system that would safely guide pilots in blind landings, he initially intended to automate the process, either by routing coordinates from the ground to an onboard radar set or by producing a beam for the plane to detect. Before long, however, he modified his idea to put human beings at the center, finding that men were more reliable than machines.

At first, he was obliged to rethink his plans because the radar prototype that he needed—the XT-1—wouldn't be available for months. In the meantime, he decided to make a proof of concept, a "talkdown" system that could relay instructions to a pilot by voice, on the assumption that he would swap out each piece for a mechanical equivalent as soon as possible.

Alvarez and his colleagues started with the most abstract version of all. In late 1941, to establish the effectiveness of verbal directions, they drew a chalk line on the floor of a hangar at Logan Airport. Blindfolding a volunteer, they handed him a pair of headphones and told him to obey the voice in his ear. A controller on the other end guided him to the line, telling him how to walk along a straight path.

The next step was to try it in the air. Mounting a light on the undercarriage of an airplane, they filmed it with a camera attached to an XT-1 antenna, which showed that the tracking was even and steady. In January 1942, they conducted landing tests at East Boston Airport, followed by trials at Quonset Point in Rhode Island, with a pilot, Bruce Griffin, in an amphibious biplane.

Since the XT-1 was still undergoing evaluation, Larry Johnston devised a way to proceed without it. To guide a plane, they needed its range, elevation, and azimuth, or its angular distance from the north. For range, they used a simple radar set. Elevation and azimuth were taken by a pair of observers with surveyor's telescopes. Inside a truck, two human "followers" entered the coordinates into a director, an analog computer that calculated the approach path from records of actual landings.

This information was routed to a controller, usually Alvarez, who watched a bank of error meters that indicated whether the plane was off course. He maintained radio contact with Griffin, who flew "under the hood," or without visual reference to the outside world. Later tests would be conducted with a physical hood that went over the cockpit, or by soaping up the windshield, but for now, the pilot simply slouched down to avoid looking out the window.

To help Alvarez understand the information that a pilot needed, Griffin gave him lessons in a flight simulator. When Griffin complimented him on how quickly he learned to fly on instruments, he explained that operating the cyclotron had taught him to compensate for lag time. Both agreed that the talkdown method was a communication problem, not just a technical one, that called for the controller to deliver simple instructions in a manner that pilots would trust.

They arrived at a procedure that steered the plane toward the glide path, or the line of descent, at four miles from the runway. Looking at the meters, Alvarez advised Griffin by radio: "You are two miles from the airport. Go up ten yards—go twenty yards to your left. Straighten out—five. You're on the path." They were confident, Alvarez recalled, that they would soon replace the human observers with an automated system. He added ruefully, "What a shock lay in store for us."

In May, they headed for Naval Air Station Oceana in Virginia for the first trials with the XT-1. Initially, it seemed fine, but as the plane neared the runway, the radar turret began to bob violently up and down. Part of the wide beam, they realized, was hitting the earth, which acted like a mirror, bouncing it off the airplane at a misleading angle. As a result, the system thought the target was underground, with the antenna locking alternately onto the plane and its illusory image. In records of previous tests, the prototype had seemed to track until landing, but it really stopped working within three degrees of the horizon.

Alvarez blamed the XT-1 team "for leading him astray," but he was also furious with himself. The apparent availability of a viable radar system had made blind landing seem both achievable and important, and he had missed the "trouble" that Barton Hoag had told him never to overlook. At that point, it seemed easiest to cut his losses, but he had fallen in love with the talkdown approach. Unlike an automated system, it wouldn't weigh down a plane with specialized gear, and the ground setup could be as heavy or complex as necessary.

In June, Alvarez dined with Loomis and his wife, Ellen, in a suite at the Ritz-Carlton. Loomis agreed that the talkdown technique was the only way to enable blind landings in time to be used in the war. "I don't want you to go home tonight until we're both satisfied that you've come up with a design that will do the job."

Alvarez was galvanized by his trust. Thinking out loud, he speculated that a narrow beam would avoid ground reflections. To follow a plane, it had to cover a large region of sky, which it normally achieved by moving in a raster pattern, like a reader's eyes running across a page of text. But there was no known way for an antenna to scan the area quickly enough to track an aircraft.

Working together, they came up with the notion of synchronizing two separate antennas, mounted at ninety degrees and rocked mechanically. Each would sweep the sky—one vertically, the other horizontally—with a beam shaped like a beaver's tail. Fed into a single transmitter, the combination would pick up a tracking signal once every couple of seconds.

At midnight, Alvarez was allowed to leave the hotel. The experience left him forever grateful to Loomis, whom he came to see as his "other father." It also meant developing a completely new radar system, which he never would have attempted if he had known from the beginning that it would be necessary. To simplify it, he dropped the idea of automatic tracking, in favor of using "human servos"—once seen as a temporary work-around—to feed data into the director.

After Loomis authorized a Mark I prototype, Alvarez collaborated on the design with Johnston—who was credited as one of its inventors—and thirty-six engineers. By the fall, they had a version that could be parked near a runway. A transmitter truck with its own radar specialist carried the azimuth, elevation, and search antennas, along with a generator that allowed it to operate anywhere. Leveled by hydraulic jacks, it was hooked up to a control van with thick cables.

Inside the van, three "human servos" sat before the cathode ray displays that received coordinates from the antennas. Each tracker twisted a knob that moved a bright line onscreen, chasing the blip of an approaching plane so that it was covered by the cursor, in a kind of primitive video game. It routed information to the hardware used by Alvarez—the controller—to issue instructions to pilots, assisted by two radar operators who guided incoming planes out of a holding pattern.

Building it pushed the engineering team to its limits. One of its members, Charles A. Fowler, was equally struck by Alvarez's determination—anyone who hindered him "was in for one hell of a fight"—and his habit of thinking aloud. Alvarez had so many ideas that his associates learned to ignore whatever he said until he repeated it three times. At that point, they would focus on it relentlessly.

His mind was always working. As they rode to the airport one morning, Alvarez suddenly said, "Can you prove all animals jump the same height?" When Fowler asked what he meant, Alvarez observed that a man, horse, dog, and flea could all jump about six feet in the air. Their weight increased as the cube of their linear dimensions, while their muscular strength was proportional to the square—a constant relationship, regardless of size. "Ergo, all animals can jump the same height."

Alvarez could also be fallible, especially with his own brainstorms. When the control van failed to start one day, he decided to give it a nudge. "The bumpers on the radar truck were not the same height," Fowler recalled, "so Luie announced that we'd use the station wagon as a 'height transformer,' because its bumpers were midway between those of the fifteen-ton radar truck and the eight-ton operator truck. I can still remember the sickening sound as the front end of the wagon crumpled."

On December 22, 1942, Bruce Griffin soaped up his windshield and took off for the first completely blind landing at Quonset Point. One week later, while the airport was closed by a snowstorm, the team learned that an approaching group of amphibious planes was low on fuel. Working with the control tower to contact the pilots, who had never even heard of the system, they guided the lead plane to the glide path, with the others following in a daisy chain.

As word started to spread, the team relocated to East Boston. On February 5, 1943, with Griffin as his check pilot in a C-78 trainer, Alvarez made the first low approach "under the hood" by a civilian. As Griffin stood by to take control in case anything went wrong, Alvarez listened intently as his controller, Ben Greene, radioed instructions from the ground.

Later that month, they left for a demonstration in Washington that would decide the program's future. Technical issues forced them to postpone twice—the long drive loosened crucial connections in the equipment—and Alvarez knew that he would get just one more chance. On the final morning, he was greeting the military observers when Johnston took him aside to say that they still weren't ready.

As Alvarez stalled for an hour, the team sent a plane to pick up replacement vacuum tubes, finishing their repairs just in time. The successful demonstration led to orders for hundreds of units, as long as the project was given a new name. Until then, they had called it

Ground-Controlled Landing, but it was authorized only for approach, not landing itself. Alvarez simply renamed it Ground-Controlled Approach, or GCA, which satisfied everyone.

Even as they started on the Mark II, which incorporated refinements from the Eagle bombsight, Alvarez was preparing to move on. The Manhattan Project had surpassed radar as the program that was drawing the best people. Earlier that year, he had attended a conference that Oppenheimer convened in Washington on the militarization of Los Alamos. Following Rabi's lead, Alvarez and McMillan agreed that it had to remain a civilian operation to maintain its independence.

In March, he spent a week at the 184-inch cyclotron in the Berkeley hills that Lawrence had turned into a mass spectrometer. "My old love for magnets, vacuum plumbing, and cyclotron electronics was rekindled," Alvarez remembered. A long night with Lawrence and other friends—fueled by multiple cocktails—ended in the worst hangover of his life. The next morning, John Lawrence informed him that the best cure was oxygen. Donning a mask in an altitude test chamber, Alvarez found that his headache was gone within minutes.

On April 16, 1943, Alvarez wrote to Lawrence that he wanted to return to Berkeley. "I am assuming that this will fit in with your plans for me, but I am writing this letter just to make sure, before I burn all my bridges." Lawrence responded enthusiastically: "What you and I will be doing next August is in the lap of the gods. At least I think we will be doing something, and it is a comfort to know that whatever we are up against next fall, you will be in there pitching on the team."

As Alvarez and Geraldine prepared to resume their life in California, however, other colleagues thought that he would be more useful at Los Alamos. At a long meeting with Oppenheimer one afternoon in New York, Alvarez heard for the first time about a bomb based on fusion, rather than fission. According to the theorists, the thermonuclear bomb was potentially unlimited in size, and Oppenheimer expressed no concerns over its moral implications.

Oppenheimer, who knew exactly how to appeal to Alvarez, told him that the fusion bomb was the best reason to come to New Mexico.

"Remarkably, in retrospect," Alvarez said, "he left me with the impression that the problem of making an atomic bomb was essentially solved and that the major interest at Los Alamos was the thermonuclear—the Super, as it was always called—the hydrogen bomb." No one yet anticipated the difficulties in building the plutonium bomb known as Fat Man, which would soon test all of Alvarez's ingenuity.

Alvarez agreed to sign up, as soon as he fulfilled one last obligation. On June 10, 1943, he left for England, with the Mark I—earmarked for the British—traveling separately by battleship. A month later, his team followed, sharing a plane with Bob Hope's USO troupe. "I never did find out who the four mysterious young men were," Hope later wrote. "But the preferential treatment they got when we landed in Foynes really surprised me."

In London, Alvarez could part the blackout curtains at his hotel to see German bombers lit up by searchlights, followed by red bursts of antiaircraft fire. At Elsham Wolds, a Royal Air Force station in Lincolnshire, he demonstrated the Mark I using every kind of aircraft, often flown by exhausted pilots returning from raids. Seeing the blood in damaged planes, he noted that the British scientists—the "boffins" in the small back room—never forgot the men whose lives were at stake.

He occasionally found himself in proximity to danger. One evening, he was having a drink at the bar at the station when an explosion shook the building. Going outside, he saw a column of black smoke. One of the "cookies," or cannisters of high explosives that the bombers carried, had caught fire and detonated, sending a shard of metal into a radio operator's head.

His closest call came on July 24. To attend a briefing near Cambridge, Alvarez took off with a pilot, N. P. Mingard, in a Miles Master monoplane. At cruising altitude, a loud clank was followed by an ominous silence. As smoke seeped into the cockpit, Alvarez saw that their propeller had stopped. Mingard calmly remarked, "The engine's packed up. We're coming down."

Banking to disperse the smoke, Mingard glided to an open field. As they descended, they realized that it was crossed with ridges, one

The Ground-Controlled Approach Mark I control and radar vans at Elsham Wolds, England, in the summer of 1943. From left to right on the radar van are (A) the search antenna, housed in a cylindrical plywood radome; (B) the horizontal azimuth antenna, mounted on one side of the van; and (C) the vertical elevation antenna. Photo by Marilee B. Bailey. Lawrence Berkeley National Laboratory.

of which struck a wheel as they landed. Their left wingtip hit the ground, tearing it off, and they halted after a hundred yards, losing half of the right wing as well. Alvarez emerged, unhurt, as the local boys ran up, disappointed that it wasn't a German plane. He paused only to cut a piece of canvas from the wing as a souvenir, and then they headed to base.

As they entered the briefing room, they heard an intelligence officer speak: "The target for the night is Hamburg." Watching the planes take off for Operation Gomorrah, Alvarez was interested in the first use of "chaff," the aluminum streamers that were dropped to confuse enemy radar, but the mission had a more lasting legacy for strategic warfare. Given the lack of accurate radar bombsights—the Eagle was still under development—the Allies had moved to civilian targets. Over two nights, eight square miles of Hamburg were destroyed with high explosives and incendiaries. It was a foreshadowing of horrors to come.

Alvarez was preparing to return to America. At Cavendish Lab in Cambridge, he ran into a friend, Bernard Kinsey, who was conducting fission studies for Tube Alloys, the British atomic program. Despite Alvarez's lack of clearance, Kinsey openly discussed the details. The bomb was the greatest military secret of all time, but it was widely known to everyone "in the club."

Shortly before leaving, Alvarez received a telegram with an unexpected request from Oppenheimer. In Illinois, Fermi had requested the presence of his longtime colleague Emilio Segrè. Oppenheimer wanted to keep Segrè at Los Alamos, so Alvarez's name had been floated instead. Delighted that Fermi respected his talents enough to accept him as a substitute, Alvarez gladly agreed.

On his last day in London, a cocktail reception in his honor testified to the success of GCA. A program that started as a gleam in his eye had grown to cost a staggering $100 million. Dozens of units went into the field, from Europe to the Aleutian Islands. At Iwo Jima, it allowed bombers to land in blackout conditions, and as attacks commenced on the Japanese mainland, it guided hundreds of emergency landings. It would be used by the military through the end of the decade, most famously in the fog of the Berlin Airlift, which began in 1948.

Alvarez would often be approached by strangers who told him that his blind landing system had saved their lives. If there was any irony in his departure to work on the most devastating weapon in history, he was untroubled. He had spent enough time with RAF pilots to see no distinction among the various ways that science could support an Allied victory, and he felt that he had been granted an understanding of such issues that many of his associates lacked.

As for the Mark I, it was sent to Cornwall—where the miserable weather was perfect for training—and entrusted to a young technical officer named Arthur C. Clarke. "It would not have been possible to design a more stimulating environment for a would-be science fiction writer," Clarke recalled. Looking at the faces of operators lit up by radar screens, he felt that he was living the future, and he even used the Mark I to aim pulses at the moon.

Alvarez's last day at the MIT Rad Lab in September 1943. Luis W. Alvarez family archives.

Although the set's power was too weak for him to receive a signal in return, he continued to think about radio waves and space, resulting years afterward in a revolutionary proposal to establish a communication system using geostationary satellites. And while they didn't meet until after the war, Clarke—who became one of the most famous science fiction authors of his generation—featured a fictionalized version of Alvarez in his autobiographical 1963 novel *Glide Path*.

In the book, Alvarez was "Professor Schuster," the eccentric scientist who conceived of the blind landing system. At one point, Clarke drew a comparison between Schuster and another character, leading to a prediction that turned out to be remarkably prescient, if slightly ahead of schedule: "Dr. Wendt was a highly capable engineer who would always be a leader of his profession, but Professor Schuster would be a Nobel prize-winner in the 1950's. All that separated them was the yawning, immeasurable gulf between superb talent and simple genius."

II.

TECHNICALLY SWEET

1943–1963

When you see something that is technically sweet, you go ahead and do it and you argue about what to do about it only after you have had your technical success.

—J. Robert Oppenheimer

4.

THE FIREBREAK

1943–1945

IN SEPTEMBER 1943, ALVAREZ AND GERALDINE LOADED THEIR possessions into a green Chevrolet and embarked with Walt to Chicago. They would be living with Geraldine's mother in an apartment on Phillips Avenue, a short drive from the university. According to Alvarez, the move gave Geraldine "a great shot in the arm." His trip to England had separated them for three months, but although their life together would become more settled, he had no intention of easing up on his work.

On his return to Ryerson, Alvarez found that the physics department had been taken over by Arthur Compton's plutonium project, so he would be spending most of his time elsewhere. A bus ran every morning to the Argonne Forest outside Lemont, Illinois, thirty miles southwest, where he had been given a chance to work with the scientist he admired above almost all others. Much later, when asked to name the greatest physicist of the twentieth century, he thought in silence and finally said, "I have to say Einstein. But Fermi is so close."

He was given a tour of the new lab from Fermi himself. Argonne was built around the reactor building, where the atomic pile from the University of Chicago squash court had been reassembled under the code name "CP-2." In its expanded form, it was a cube thirty feet to a side, with fans and concrete shielding that allowed it to run at higher levels than before. At the top was a thermal column, a graphite "wick" from which only slow neutrons emerged.

Alvarez hoped to prove that Fermi had been justified in accepting him instead of Segrè, but he soon realized that Argonne was far from the best outlet for his gifts. Most of the important work there had already been done, and the project's focus was shifting to the construction of a reactor in Oak Ridge, Tennessee, to produce plutonium on a larger scale. This intermediate reactor, in turn, would inform the design of the industrial reactors in Hanford, Washington, which would provide most of the material for the plutonium bombs.

In theory, once the equipment was working, production would be relatively straightforward. During a controlled chain reaction, a fraction of the uranium absorbed a neutron, decaying into plutonium-239, which could be separated out with techniques developed by Glenn Seaborg. In practice, it suffered from neutron "poisoning," with impurities in the uranium and graphite capturing so many neutrons that the reaction was impossible to sustain.

Because there was no way to chemically detect these contaminants in such infinitesimal amounts, Argonne was conducting tests in a simple, if tedious, fashion, by placing manufactured materials in the reactor, running it, and comparing the resulting radioactivity to a standard baseline. It was indispensable, but dull. Alvarez quickly learned to operate the reactor, but apart from proposing a few improvements, he did whatever he could to find work elsewhere.

At the weekly committee meetings in Chicago, he was surprised to see physicists like Samuel Allison and Eugene Wigner obsessing over technical details. "I made it a point not to examine blueprints in order to second-guess my engineers," Alvarez later said. "I discussed a project at great length so that they came to appreciate the problems that I felt had to be solved, but I left them on their own to solve those problems. I operated on the belief that if engineers know that physicists are going to check their blueprints they won't be nearly so careful."

Alvarez saw the emotional costs of this kind of work firsthand. Not long after his arrival, he received a call from Lawrence, who was scaling up magnetic separation of U-235 at Oak Ridge. Without this process, there would be no uranium bomb, but it was plagued by impurities that shorted out the magnets. Stressed and overworked, Lawrence suffered from backaches and bronchial problems, and he finally checked himself into the hospital for a week in Chicago.

When Alvarez brought him some books, he found Lawrence wearing an orthopedic corset, which had to be tightened by an orderly during his visit. Trying to make conversation, Alvarez mentioned an idea for a project after the war—he wanted to build a linear accelerator using surplus radar sets, on the premise that physics funding would drop to its prewar levels. "I assume that your isotope separation plant will also be surplus," Alvarez joked. Lawrence wasn't amused—it was too soon—so Alvarez switched to a less sensitive subject.

As a side effect of the bomb project, the Argonne reactor was a veritable factory for slow neutrons, and Fermi saw no reason not to take advantage of it for his own research. A few weeks after Alvarez arrived, Fermi mentioned an experiment that he had in mind. It would yield a more accurate measurement of a standard neutron source, with a method that involved bombarding a tank of borax and magnesium salt as a reference point. Sensing that it was a kind of test, Alvarez agreed to carry it out. The tank gave him trouble, however, and after he did a bad job of soldering a spigot, a growing puddle was left on the floor.

Since he didn't have any friends in the machine shop, he had to stop the leak himself. At first, he tried to repair the tank while it was full, even after he was reminded that his acetylene torch couldn't heat the solder past the boiling point of the water inside. Alvarez was too stubborn to stop, and after two days, he somehow managed to fix it. Despite his frustration, he remembered it as a turning point. "It convinced me that I wanted to be a technical man, hands-on, not the administrator I had become who would have called up a technician."

In retrospect, he saw Chicago as "a decompression chamber" between the Rad Lab and Los Alamos—a circle of calm at the heart of the maelstrom. Until he found a way to return to defense work, he wanted to make good use of his time. Although he was a free agent, it was hard to think of projects that struck him—and Fermi—as worthwhile, while retaining the flexibility to leave at a moment's notice. Eventually, he focused on finding long-range alpha particles from bombarded uranium, which he had once missed seeing in Berkeley.

Along the way, he became closer to the other scientists, starting with the only woman on Fermi's team. Twenty-four, newly married, and pregnant, Leona Woods Marshall was building a velocity selector with a rotating "chopper" that let through neutrons at specific speeds, unlike Alvarez's slow neutron beam, which simply ignored the unwanted energies. One day, she was assembling a circuit on top of the pile when Alvarez showed up and said, "Give me another soldering iron and let's go at it."

At the University of Chicago, she had heard about Alvarez, who "was high on my list of mythological figures, above griffins and slightly below Michelson and Morley." Afterward, he praised Marshall, who saw him as a friend and advisor, as "certainly the finest woman physicist our country has ever produced," although he privately felt that she "did several great things with Fermi after the war, but nothing of note either before or after that partnership." In fact, her richly varied interests—from cosmic rays to paleoclimate studies—hinted at what Alvarez might have been like if he had been obliged to navigate the scientific world as a woman.

Alvarez was more invested in his relationship with Leslie Groves, the heavyset brigadier general who oversaw the atomic bomb program. Not long after Alvarez's arrival in Chicago, Groves summoned him to the district manager's office, explaining that Lawrence had recommended him for a project. Groves wanted to find any nuclear reactors that had been built by the Nazis, which could then be targeted for destruction by air. After giving him a week to look into it, Groves dismissed him with a warning not to tell anyone else.

Understanding the importance of the assignment, Alvarez returned with a plan based on xenon-133, a radioactive gas emitted by nuclear reactors, which could be collected from the atmosphere by a "scrubber" in a plane. After Groves signed off on the idea, Alvarez consulted with General Electric—its light bulb business made it a useful source of information about noble gases—and designed gear for a test plane in Ohio that "scared a lot of cows."

The scrubber was installed in three Douglas A-26 bombers that flew reconnaissance missions over Germany. Since the Nazis never produced a chain reaction—removing a fundamental justification for the Manhattan Project itself—the planes detected nothing. The program's most notable outcome was Groves's partnership with Alvarez, who exhibited the same tenacity that the general admired in Lawrence. He evidently liked Alvarez's radar background, and they occasionally conferred in the waiting room of Union Station in Chicago, where Groves received a line of visitors on stopovers during his railway trips across the country.

Shortly afterward, Groves named Alvarez to the group that would fly to any location where the Axis had dropped an atomic bomb, with a plane with test equipment kept constantly at the ready. Alvarez also consulted on the Alsos Mission, which gathered intelligence on Nazi scientific programs. He worked with agents who were looking for traces of heavy water in Switzerland, advising them to sample the swiftest stretches of the Rhine and areas of Lake Constance that were fed by German rivers. Finally, he agreed to quietly give Groves updates on Los Alamos, since Oppenheimer was inclined to keep his plans to himself.

Alvarez's strongest memories from this period were of Fermi, who ate lunch with him in the cafeteria nearly every day for six months. The forty-two-year-old Italian immigrant was charming and unassuming, but Alvarez saw that he had no false modesty: "I'm sure that if I had asked him who in his opinion was the best physicist alive he would have thought for perhaps half a minute and nominated himself."

At parties at Fermi's house, Alvarez was introduced to table soccer, or foosball, although he never managed to score against his hero. Fermi also taught him an Italian game in which two players extended fingers at the same time, alternating rounds to get as close to fifteen points as possible. Alvarez improved until they were both equally good, and he did even better with a game they played on graph paper, trying to connect five Xs or Os in a row: "It really made [Fermi] feel quite bad, that he wasn't able to do as well as I did."

He rarely had any other reason to feel superior to "the only completely self-sufficient physicist I've ever met." Fermi had taught himself to oversee huge experimental projects, but he could also work perfectly well with just half a dozen reference books, along with his own carefully maintained journals. The inescapable conclusion was that he didn't really need Alvarez, who was unlikely to contribute anything that Fermi couldn't do without him.

One day, Fermi mentioned over lunch that he was wondering if neutrons could be reflected. He suspected that they would behave like X-rays, which bounced off a solid if they struck it at an extremely shallow angle. Inviting his colleagues to his office to discuss it further, he noted that to calculate the angle in question, they could use the standard formula for the refractive index, or the amount that radiation was bent by a given material.

When Fermi couldn't remember the equation offhand, Alvarez told him that it was in a book that he could retrieve from next door. Before he could leave, Fermi told him that it wouldn't be necessary. Going to the blackboard, he wrote out Maxwell's equations—the fundamentals of classical electromagnetism—and plugged in the right variable to account for refraction.

As the others watched with amazement, Fermi derived the formula from scratch, unhurriedly writing out each step, as if he were copying it from a textbook. When he was done, he checked his pocket slide rule, frowned, and said that the angle would be too small to measure. Thinking fast, Alvarez came up with a reason that it might be larger, drawing on his knowledge of the literature. He proudly recalled, "Enrico immediately perked up and said I was right."

It was one of the few times that Alvarez told Fermi something that he didn't know. "That night at home," Alvarez remembered, "I reproduced [the equation] and was quite pleased with myself. If one step was easy enough to allow me to go faster than he did, the next was so difficult that I could never have managed it alone. But Enrico worked the difficult steps at exactly the same rate he worked the easy ones." Years later, he said admiringly, "There is no democracy in physics. We can't say that some second-rate guy has as much right to opinion as Fermi."

Fermi's intellectual independence sometimes led him to keep his own counsel. In late 1943, William Havens and James Rainwater at

Columbia reported that they had done something that Alvarez had never quite managed to pull off—they used the time-of-flight method to calculate the energy levels at which a neutron was more likely to be captured. For obvious reasons, they tried it immediately on U-235 and U-238, delivering the results without any fanfare. "I was with Fermi when he saw them," Alvarez wrote. "He was absolutely baffled."

The curve for U-238 looked exactly as expected, with bumps at energies where neutron absorption was more frequent. It was the chart for U-235—the fissile isotope—that shocked Fermi, because it seemed more or less the same as the other one. Everyone had assumed that fission and neutron capture would be mutually exclusive, but the reactions were actually competing. Concerned that this would also be true of plutonium, Fermi secretly asked Segrè to arrange a round of tests, which would upend the entire Manhattan Project.

Alvarez was starting to realize that if he wanted to make an impact, he had to detach himself from Fermi. On March 25, 1944, he received a cable from Oppenheimer: "Shall be in Chicago Thursday and Friday at least. Looking forward very much to seeing you probably Friday. Have urgent news for you." On the copy of the telegram that he kept in his private files, Alvarez added a handwritten annotation: "Second invitation to Los Alamos."

On a visit from Robert Bacher, the head of the Experimental Physics Division, Alvarez affirmed that he was ready to leave. Bacher, who had worked with him on radar, already had something in mind. The uranium and plutonium bombs were both expected to utilize the gun assembly method, in which one piece of fissile material was fired at another to start a chain reaction. An alternative approach used explosives to implode a plutonium shell into a solid sphere. It was seen as a backup, but Bacher wanted Alvarez on the implosion team.

Alvarez gladly agreed, even though it would mean uprooting his family yet again. After quickly finishing up his work on long-range alpha particles, he left Chicago by train in April. On his way down, he shared a drawing room compartment with Edward Teller, another one of his heroes. Teller was warm and friendly, but there was little chance for small talk, as the Hungarian physicist spent the trip giving Alvarez a crash course on the dynamics of implosion.

He was surprised by the beauty of Los Alamos. A drive on a steep, winding road led to a high mesa covered in pine forest and scrub. Past the gates manned by military police, who guarded the only entrances through the barbed wire, he found himself in a secret town of about four thousand residents—mostly young, white, and male—who were trying to maintain a semblance of normality. On the day he arrived, they were holding a tennis tournament.

Alvarez was given a room in the rustic Fuller Lodge, the dining hall of the boys' school that had once occupied the property, next to George Kistiakowsky, the head of the implosion group. His first order of business was to find a suitable living arrangement for his wife and son, who were scheduled to follow him in a month. Geraldine was pregnant, qualifying them for a larger apartment, and at the housing office, Vera Williams, the wife of an old friend from Chicago, promised that she would reserve the best one that became available.

Most of the houses were in units of four, with two apartments on each level. Williams later said that when the other tenants learned that their new neighbors would be "the Alvarezes," they associated the name with the locals who did the cleaning and maintenance work. After they complained "of their imminent loss of social standing," Williams set their minds at ease. Alvarez found that the previous resident had abruptly moved into the bachelor officers' quarters—his marriage had ended, which wasn't unusual at Los Alamos—and left "personal belongings, dirty clothes, uneaten food, and the worst mess I have ever seen in my life."

Although he cleaned it up before Geraldine and Walt arrived, he would rarely have time for his family. He was put to work at once by Oppenheimer, whose transformation filled him with awe. Alvarez suspected that Groves had named Oppie as director on Lawrence's recommendation: "If you ever saw Robert Oppenheimer in those days with the fuzzy hair and the wild-eyed look, it's not the kind of a guy that a West Point general would want to put in charge."

Alvarez's official badge photo at Los Alamos. Los Alamos National Laboratory.

Confounding his doubters, Oppenheimer turned out to be more than Lawrence or Alvarez's equal as a team leader—and on a vastly greater scale. Alvarez later wrote of Oppenheimer, "His haircut almost as short as a military officer's, he ran an organization of thousands, including some of the best theoretical and experimental physicists and engineers in the world. The laboratory's fantastic morale could be traced directly to the personal quality of Robert's guidance."

Because he had to delegate based on limited information, one of Oppenheimer's great strengths was his intuitive sense of where a scientist would be the most valuable. Looking at Alvarez's résumé, he appointed him to a steering committee for the governance of the lab. He also asked him to run a project to evaluate the implosion method. Because it involved detecting radiolanthanum inside an imploding shell, in a kind of X-ray in reverse, it was known as the RaLa experiment.

Alvarez was more interested in assisting Kistiakowsky, the versatile and headstrong explosives expert who had once worked in Alfred

Loomis's private lab in Tuxedo Park. Every day, Alvarez bicycled to his office in the Technical Area, where he and Kistiakowsky sat at facing desks. At first, his role was to serve as the implosion chief's "eyes and ears," updating him on developments connected to their work, which was about to change in ways that no one had foreseen.

For months, Segrè had been in a remote cabin, studying spontaneous fission—the unprompted production of secondary neutrons—in plutonium. The project, which was part of the reason that Oppenheimer had kept him from joining Fermi, initially seemed like a waste of his talents. When Segrè checked samples from the Berkeley cyclotron, the numbers all seemed fine.

Turning to plutonium produced in an atomic pile, however, he found something troubling. Its spontaneous fission rate was much higher than in material from the cyclotron, making it more likely that a stray neutron would start a chain reaction before the assembly was complete. As heat caused the core to expand, it would blow apart too soon—the "fizzle" known as predetonation. The same would be true of Pu-239 from the production reactors at Hanford, which were expected to supply the plutonium needed for more than one bomb.

It was Fermi who finally understood what was happening. In the reactors, Pu-239 was capturing neutrons to become Pu-240 at a much greater rate than in the cyclotron. Fission and capture were competing, just as he had seen in the uranium curves from Havens and Rainwater. And because Pu-240 had a higher rate of spontaneous fission, it would lead to predetonation in any bomb using the gun method. The plutonium "bullet" would simply melt.

Alvarez later said that the realization that a plutonium gun would be too slow to produce a bomb "was the most important single event at Los Alamos in the first months after I arrived." Incredibly enough, Fermi had been aware of the possibility before Alvarez left Chicago, but he had said nothing to his associates at Argonne. According to one history of the Manhattan Project, Segrè speculated that it was a case of wishful thinking: "Fermi, like Groves, might consciously or unconsciously have been wishing the results would go away."

In all likelihood, Fermi hoped that it would turn out to be a minor distraction, so he decided not to reveal it prematurely, withholding the truth for the greater good of the project. Alvarez might well have approved of this decision, but now there was no overlooking the problem, which moved his group from the fringe to the center. Kistiakowsky and Alvarez had assumed that the implosion method would remain a fallback option, but it soon became clear that it was the only way that a second bomb would be ready before the end of the war.

ALONG WITH THROWING HIMSELF into his official assignments from Oppenheimer, Alvarez resolved to learn everything about implosion. He devoured reports in the library, took part in seminars, and attended meetings of the different divisions. Technical details were openly discussed, regardless of secrecy, which he saw as essential to success: "I'm sure the weekly colloquia provided Klaus Fuchs"—a theoretical physicist who was studying implosion as well—"with much of the information he passed on to Moscow. I'm no less sure that they also saved countless American and Japanese lives."

Although no one had asked him to master this material, it seemed like the only way to make a meaningful contribution. The entire Manhattan Project was based on the problem of assembling a critical mass—the amount of fissile material that would sustain an explosive chain reaction—while avoiding the stray neutrons that would start the process prematurely. "This problem is like running from the front door to an automobile in the rain," Alvarez wrote. "If getting as much as a single drop on your head is a very serious matter, in a light drizzle you run as fast as possible."

After Segrè's plutonium tests revealed "a rainstorm of neutrons," the focus shifted to implosion, which could assemble the masses more quickly than the gun method used in the uranium bomb known as Little Boy. Eventually, the alternative design called Fat Man ended up in the hands of Kistiakowsky, who was still skeptical of the approach when

Alvarez arrived: "He felt Dr. Oppenheimer was mad, almost, to think that such an absurd object could ever be made to work."

In time, they came to believe in it, aided by the mathematicians John von Neumann and Stanisław Ulam, who were calculating the dynamics of the explosive lenses that would fit together over the bomb's core. Much like an optical lens, the precisely shaped charges could focus the shock waves to converge at the center, avoiding the disruptive jets of liquid metal produced by collisions between wavefronts—but only if the process was perfectly symmetrical.

They came up with a pattern of thirty-two outer lenses, twenty hexagonal and twelve pentagonal, like the panels of a soccer ball. Each lens was cast from two different explosive compounds that combined to make a converging spherical wave. They were set off with Primacord, a plastic tube filled with powdered explosive, with a separate fuse running from each lens to a shared detonator. The bulky result was hard to fit into a bomb case, and because the cord's density was uneven, there was no guarantee that it would all fire simultaneously.

Toward the end of May 1944, Alvarez asked Kistiakowsky why they hadn't taken the opposite approach, by wiring thirty-two electric detonators—one for each lens—to go off at the same time. Kistiakowsky told him that the timing there was even worse, since the spread between detonators could vary by half a millisecond. Looking into the problem, Alvarez saw that the culprit was the explosive primer in the detonators, which was ignited by a bridgewire. It struck him that it should be possible to eliminate this intermediate step. With a powerful capacitor, a device for discharging electrical energy, they could just blow up the wire itself.

Kistiakowsky seemed doubtful, but he replied, "We've got to try anything, so you go ahead and give it a shot." Although he was a newcomer to the subject, Alvarez felt that his inexperience might be as much of an advantage here as it had been in radar. "By this time I had learned just enough about high explosives to experiment with them but not enough to be sure that an idea wouldn't work."

First he had to prove that high explosives could be set off directly by blowing up a bridgewire. Since he was busy enough as it was, Alvarez enlisted his friend Larry Johnston, a recent arrival at Los Alamos. On a remote area of South Mesa, Johnston opened a standard detonator, poured out the primer powder, and replaced it with a more sensitive explosive. Wiring it to a capacitor, he took shelter in a portable hut and discharged it. The satisfying bang—and the copper shards driven into the other side of the wall—indicated that the idea had potential.

Kistiakowsky was still unsure that it could trigger multiple detonators at precisely the same time. In another test, Johnston taped detonators to either end of a lead plate. When set off together, their shock waves would collide somewhere in the middle, leaving a visible crease. By measuring its distance from the halfway point, he could tell how far apart the detonations had been. With exploding bridgewires, the crease was just an eighth of an inch off-center, meaning that they detonated within a single microsecond—a thousand times better than the usual method.

At a meeting that summer, Oppenheimer signed off on the program: "We'll proceed with Luie's approach." It was yet another idea that Alvarez had developed in private that suddenly put him in charge of a large team. He ordered tens of thousands of blasting caps from Pasadena, learning to specify the most expensive materials available, while Johnston was joined on the mesa by enlisted men with backgrounds in engineering. To produce enough detonators in time, Tewa women were bused from nearby pueblos to solder the bridgewires and load the powder, working behind shields in case of an accidental explosion.

Because it eliminated the need for a primer, an exploding bridgewire was safer than other methods, but countless problems had to be solved. To minimize the spread between detonations, they tried everything imaginable—changing the wire material, packing the explosive in different ways, even sorting it by particle size. The timing was checked with a camera that recorded the flash of the explosion on film. When the detonators were triggered in a vertical column, any deviations would show up as displacements of the points of light from a straight line.

Alvarez and Johnston's group ultimately reduced the spread to a few hundredths of a microsecond, establishing a new standard for high explosives. Feeling that he had done all that he could, Alvarez passed the project along to the team that would figure out the engineering details, which continued almost to the day of the Trinity test. Explosive lenses and precision detonators were the two innovations that made the plutonium bomb possible, ensuring symmetry in both space and time, and neither would have worked without the other.

On most days, Alvarez was more involved with the RaLa experiment, which Oppenheimer saw as the perfect fit for his skills. The physicist Robert Serber had proposed testing the implosion method by placing a gamma ray source inside a spherical metal shell. When it was compressed by explosives, the walls increased in thickness, absorbing more radiation. An array of nearby ionization chambers would register the drop in radioactivity in the fraction of a second before they were destroyed by the blast, resulting in a series of diagnostic snapshots of the contracting sphere.

To provide the gamma rays, Alvarez arranged for radiobarium to be brought in a lead cask from the reactors at Oak Ridge, carried in a truck that drove nonstop until it reached its destination. It was the most radioactive material that any of them had ever handled. Since it had to be "milked" to yield radiolanthanum, Alvarez designed a mobile separation facility, along with a kind of fishing rod used to insert the sample into the shell from a distance.

Testing took place in Bayo Canyon, two miles north of Los Alamos. To protect the recording equipment that received the data, Alvarez found two tanks at Dugway Proving Ground in Utah. With their guns removed, they were large enough to serve as field labs, which he and Kistiakowsky appraised firsthand in July. Climbing into one tank, Alvarez watched through the periscope as hundreds of pounds of explosive blew a wave of dust across his field of vision. When they emerged, the team noticed fires sprouting around them—fragments of hot metal had ignited the brush—and barely managed to put out the blaze.

The first real detonation was on September 22, 1944, followed by two more the following month. One day, while they were handling the radiolanthanum, they accidentally broke one of the wires on which it was suspended, leaving it dangling down on one side. A health protection officer warned them to stay ten feet away, so they spent two hours trying to fix it using improvised tools with long handles, exposing themselves to radiation the entire time.

Finally, one man said, "I can go in there with long-nose pliers and be out before I get any dose at all." After arguing with the health officer, the volunteer ran up to the sample, fixed the wire, and retreated after taking nearly a full day's worth of radiation. The test could proceed, although, as a project history noted, "by today's standards, the exposures they received would be judged unacceptable."

For months, they failed to achieve uniform compression. The breakthrough occurred after they shifted from the shell to a solid sphere that wasn't dense enough for a critical mass. When a plutonium "pit" was squeezed by explosives, its density would increase enough to start a chain reaction. By the end of the year, this approach—combined with Alvarez's detonators—yielded a workable design. Oppenheimer concluded, "Now we have our bomb."

Over the previous six months, Oppenheimer had overhauled Los Alamos to focus on implosion, which Alvarez saw as the director's greatest achievement: "Most people seem unaware that if separated U-235 is at hand it's a trivial job to set off a nuclear explosion, whereas if only plutonium is available, making it explode is the most difficult technical job I know." By focusing intently on the problems that it presented, Alvarez ended up closer to the center of it than he could have hoped.

He relished the chance to hold his own with other physicists, ranging from idols like Niels Bohr to younger associates like Richard Feynman, who shared his rebellious streak. Feynman made a game of cracking the combination locks on the office file cabinets, telling Alvarez that it was easy if you knew how. Taking apart his own lock, Alvarez learned how to find the numbers for any open safe by slowly turning the wheel and examining the movement of the bolt. Unlike Feynman, he treated the solution as its own reward, and he never put his knowledge to use.

Alvarez became friendly with another group of colleagues at a weekly poker game. John von Neumann was widely seen as the brightest man at Los Alamos, but Alvarez noted that his mastery of game theory—a field that he invented almost single-handedly—was less than useful under such circumstances. On most nights, he wrote, von Neumann "left our poker games poorer than when he arrived."

Within the meritocracy of the lab, Alvarez looked for ways to stand out. The historian Richard Rhodes said, "People who I talked to at Los Alamos really did not like him very much. He was the kind of guy who would walk into the room if you were playing pool and say, 'I can beat you at this game.'" True to his word, Alvarez won first place in the Manhattan Project billiards tournament—Feynman was another entrant—and the youngest physicist on the mesa, Ted Hall, suffered this treatment firsthand: "Luis Alvarez used to kill me on the billiards table."

Alvarez's casual attitude toward security occasionally raised red flags. In the fall of 1944, a source said that Alvarez "was allegedly highly informative concerning his work in letters to his relatives," presumably before his arrival at Los Alamos, where any sensitive disclosures would have been caught by the censors. Rumors about his wartime activities had already spread from Minnesota to Berkeley, but no further investigation was pursued.

Despite these issues, Alvarez earned a position of trust. On a Pentagon panel chaired by the mathematician Theodore von Kármán, which met on a monthly basis to discuss science's role in the postwar world, he served as a radar expert. He also stayed in touch with Groves, circumventing official lines of communication to pass along what he knew about Oppenheimer's intentions.

On one occasion, it backfired. After Groves mentioned a piece of precision equipment that needed to be built, Alvarez forwarded the request to the machine shop in Chicago. When Oppenheimer saw the apparatus there, he grew pale, recognizing it as "secret inside stuff" that indicated a possible leak. Realizing that he had exceeded his authority, Alvarez explained what he had done, but avoided mentioning Groves.

He wanted to stay on good terms with both men, although the author Nuel Pharr Davis later wrote that he had aroused the director's suspicions: "Oppenheimer could never be sure what to expect from Alvarez."

All the while, Geraldine—like most wives at Los Alamos—was left in the dark about the details of his work. To make it easier for her to keep house, Alvarez arranged to install an electric oven instead of the standard wood-burning stove. The units in their building shared a furnace that was always too hot, and since there wasn't any way to adjust it, they learned to leave the windows open.

On October 3, 1944, exactly four years after Walt was born, Geraldine gave birth to Jean Alvarez—who shared her name with the baby they had lost—at the Los Alamos hospital. Geraldine grew close to the other young mothers, with one notable exception, as Jean later recounted: "The impression I had from my mom was that no one connected very well with Kitty Oppenheimer."

Whenever he had time, Alvarez went on evening walks with his son, showing him the buildings under construction, and visited Santa Fe or the Jemez Mountains with his family. In the winter, he skied on a slope that Kistiakowsky and his colleague Hugh Bradner had cleared of trees with plastic explosive, returning for outings with Geraldine and the children in the canyon below. Jean recalled that Geraldine came to love the geography of the Southwest, as well as Pueblo culture, which provided an outlet for her natural curiosity: "I think she essentially enjoyed Los Alamos."

To relieve the tension, wild parties were held every weekend, with punch spiked with ethyl alcohol from the lab. One was hosted by Kenneth Bainbridge, the physicist overseeing the Trinity test, in honor of I. I. Rabi, who had just won the Nobel Prize. "Geraldine, who was normally the model of decorous behavior, for some reason bet Ken Bainbridge a substantial amount of money that she would let herself be dunked in his bathtub," Alvarez said. "She went home, put on her bathing suit, came back, and in full view of all who could cram into the bathroom, won the bet."

In private, however, the stress was hard on them both. During an outbreak of hepatitis in November, Alvarez spent two weeks in the hospital, which sidelined him at the worst possible time. The isolation of Los Alamos strained many marriages, and his inability to discuss what he was doing produced a growing sense of distance: "Gerry and I forgot how to share our lives."

He would soon be taken even farther away. In the spring of 1945, with the implosion project in the hands of others, Alvarez decided that he wanted to be there when the bomb was used. When he asked for an assignment in the Pacific, Oppenheimer allegedly said, "As a matter of fact, Luis, I have just a job for you, and you won't have very much time to get it going. I just realized that we have no way to measure the energy released in the bombs that we are going to drop over Japan."

Alvarez understood immediately. Although they were on track to produce at least two plutonium bombs, one of which would be tested at Alamogordo, there was enough uranium for only one more weapon. Since it had to be saved for the Japanese, there wouldn't be a chance to observe a blast under the controlled conditions of a proving ground. Oppenheimer wanted to find a reliable method that could be used to measure it over enemy territory.

In reality, the idea may have originated with Alvarez, who was looking for a reason to participate in the bombing. Realizing that the only factor that could be detected from the air was blast pressure, which he could treat as a sound wave, he remembered a report that he had read while waiting in an office in Pasadena. To train aircraft gunners, physicists at Caltech had developed a pair of microphones that could pick up shock waves from bullets. The signals could be used to calculate the distance and angle of the shots—or, in theory, the explosive energy of a bomb.

Alvarez recruited one of the inventors, Wolfgang "Pief" Panofsky, who joined a team that included Larry Johnston, Bernard Waldman, and Harold Agnew. At Wendover Field in Utah, they went on a practice run with the 509th Composite Group, the unit that would be responsible for the atomic bombings. As a B-29 dropped a dummy bomb over the Salton Sea, they sketched the rear compartment, trying to figure out how to install an airborne lab in the limited space available.

The group came up with a pressure gauge in an aluminum cylinder, about three feet long, that could be released by parachute from an observation plane. Inside the case were two microphones, a battery pack, and an amplifier to transmit the signals. On the airplane, a receiver for each gauge would be hooked up to an oscilloscope, which could be filmed with a 16 mm camera for later analysis. For now, the plan was for Alvarez's team to hand off the equipment to a technical crew, but he privately intended to be on board himself.

A series of practice runs culminated in a full dress rehearsal in May—the day before the war ended in Europe—in which a hundred tons of conventional explosives were set off in Alamogordo. The gauges were successfully dropped from fifteen thousand feet, but just one captured the shock wave, and the numbers seemed around forty percent too low. Alvarez had to quickly make improvements, while also trying to anticipate the risks of flying near a nuclear explosion. Arranging for his plane to be tracked by radar from the control room, he was confident that he would have time to fly to safety before the decisive moment of the Trinity test.

ON THE AFTERNOON OF JULY 15, 1945, Alvarez told Geraldine that she might see "something interesting" if she looked to the south at five in the morning. Driving to Albuquerque, he had dinner at the Hilton Hotel with Lawrence; the chemist Charles Thomas; and the *New York Times* reporter William Laurence, who was preparing a history of the Manhattan Project. From there, he went to Kirtland Air Force Base, where he learned that Oppenheimer was trying to reach him.

At a public phone booth at the airport, Alvarez listened, disbelieving, as Oppenheimer informed him that the plan had changed. The B-29 would stay at least twenty-five miles away from the bomb, which meant that they couldn't drop the gauges at all. Alvarez was infuriated by what he saw as an overblown concern for safety, although the decision may have been equally motivated by the weather—a thunderstorm had hit just hours before the scheduled test, and poor visibility would make it difficult for the plane to move into position.

In any case, he had no choice but to agree. Early in the morning of July 16, they took off for Alamogordo. Alvarez positioned himself in the cockpit, kneeling between the pilot and copilot, along with William "Deak" Parsons, the head of the Ordnance Division. Panofsky and the rest of the observation team huddled in the rear, wondering if the bomb would even work. Alvarez had often reminisced about how many times his blind landing system had failed, and there were countless ways that the plutonium bomb could still go wrong.

He listened over his headphones to the countdown, which ended at 5:29 a.m. At the last second, clouds blocked their view of the tower with the bomb—a seeming annoyance that turned out to be a stroke of luck. They were wearing polarized glasses that could be darkened with a knob, which didn't protect them from the infrared spectrum. "Had we

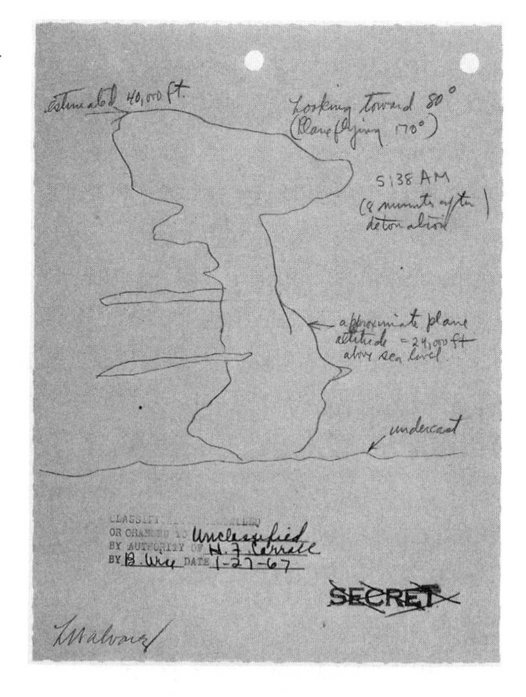

Alvarez's sketch of the mushroom cloud at the Trinity nuclear test. National Archives.

been looking directly at ground zero," Alvarez said, "we certainly would have burned holes in our retinas."

A sudden flash of light filled his field of vision for half a second. After it faded, it was followed by an orange glow, nearly as bright, as a ball of fire broke through the clouds. "This fire ball seemed to have a rough texture with irregular black lines dividing the surface of the sphere into a large number of small patches of reddish orange," Alvarez wrote in his report. It was replaced by a gigantic cloud that reminded him of a parachute blown up by a fan, above a narrower column of dust, leading him to make the first known use of a term that would soon become infamous: "This had very much the appearance of a large mushroom."

Alvarez realized that he had missed feeling the shock wave. He felt strangely detached from the entire experience, with none of the drama reported by observers on the ground. Finding that no one had thought to bring a camera, he quickly made two pencil drawings of the mushroom cloud, sketching how it looked eight and twelve minutes after the blast. He continued to watch intently as it flattened out and broke up, drifting slowly from east to west.

Far below, in the desert, Fermi prepared his own test, which was also based on blast pressure. After the flash, he took some paper scraps from his pocket, releasing them just before the shock wave hit. Using his shoes, he measured how far they were blown—about two and a half meters—and calculated that the explosive power corresponded to ten thousand tons of TNT. He was around forty percent below the best early estimates, or roughly as good as the gauges had been so far.

Alvarez felt vindicated by the knowledge that his detonators had functioned perfectly, but it was overshadowed by his awareness of the task to come, as well as his lingering anger toward Oppenheimer. He later granted that the director had been under enormous pressure, but at the time, he saw it as a betrayal. With no way of testing the gauges until they were used in Japan, Alvarez didn't know if he could ever trust Oppenheimer again: "I was absolutely furious, angry with him as I have never been angry with anyone before or since."

ACCORDING TO ALVAREZ, Oppenheimer kept a chart on the wall of his office, showing the gradual daily increase in the total inventory of U-235. A second horizontal line marked the point at which enough fissile material would be on hand to build a working weapon. Based on the rate at which the uranium stockpile was growing, it was a simple matter to guess when the two lines would meet. The plan was to drop Little Boy on Japan on the first clear day after the necessary mass was available, which seemed likely in about two weeks.

Alvarez was officially a member of Project Alberta, the section of the Manhattan Project responsible for delivering the bombs. As head of the Aerial Observation Team, he worked feverishly with Larry Johnston and Harold Agnew to prepare for their departure for the Northern Marianas. Each of them learned how to service all the equipment involved in the gauges, with checklists for every step. They packed tool kits, spare parts, and a personal box of goods for trading—Alvarez brought a case of whiskey labeled "solvents."

On the morning of July 20, 1945, they took a bus to Albuquerque and flew from there to Wendover Field, where they exchanged their civilian clothes for uniforms. Each team member received an assimilated military rank to formalize his status. Alvarez was mildly disappointed to learn that he was only a lieutenant colonel, while Robert Serber, who was serving as a special consultant, would be designated a full colonel. All of them were issued cards stating that they were entitled to a colonel's privileges, but only if captured by the enemy.

Early the following day, they boarded a Douglas C-54. After landing briefly at Hamilton Field in California, they crossed the ocean to Oahu, taking on additional passengers near Pearl Harbor before continuing across the Pacific. A refueling stop at Johnston Atoll was followed by the long journey to the Marshall Islands. Finally came Guam, Saipan, and one last short flight to the island of Tinian. It had taken them a week to travel seven thousand miles.

On July 26, they landed at the world's largest airfield. A year earlier,

An unidentified military policeman and Alvarez with the plutonium core for Fat Man on Tinian. Los Alamos National Laboratory.

the Japanese garrison of six thousand had been annihilated, leaving a few pockets of resistance, along with a civilian prisoner camp. Within months, the airbase expanded to cover the island, with hundreds of planes lined up next to huge runways of crushed limestone. Japan had been well aware of Tinian's strategic importance as a base for attacks against the mainland, 1,500 miles away, and the United States had more than fulfilled its horrific potential.

Because Tinian was roughly the shape of Manhattan, sectors were given names like Central Park and the Village. The 509th had a separate camp near North Field, with its own quarters, mess hall, and movie theater. Alvarez saw some familiar faces, including James Nolan, the chief medical officer at Los Alamos, who had treated him for hepatitis and delivered Geraldine's baby. Because of his experience with radioisotopes, Nolan had been selected to escort the uranium projectile for Little Boy on the USS *Indianapolis*. Four days after the handoff, the heavy cruiser was sunk, killing hundreds and leaving the survivors to suffer attacks by sharks.

With the privilege of rank, Alvarez and Serber shared a private tent, while the other team members were packed twelve to a hut. The weather on most days was hot, humid, and rainy. Breakfast at seven was followed by a drive past a military cemetery to the Technical Area, where guards and barbed wire surrounded two Quonset huts, one for Little Boy, the other for Fat Man. Alvarez worked in the latter, near the shelf on which the plutonium core rested in a magnesium field case. The world's most valuable material was concentrated in a sphere less than four inches across, weighing around fourteen pounds, that felt warm when he cradled it in his hands.

Setting up their benches, his team calibrated the equipment, trying to think of everything that could go wrong. After a break in the afternoon, they resumed work until early evening, when they relaxed over dinner, a movie or live show, and drinks at the officers' club. Every few days, they watched as bombers took off for night raids on runways that ended at a cliff above the ocean, disappearing for a few seconds before coming back into view. Each plane was so heavy with bombs or incendiaries that some were expected to crash into the water, which nonetheless led to fewer casualties than would result from flying more aircraft.

Alvarez was still determined to be on the observation plane that accompanied Little Boy. He shrewdly waited until he was on Tinian to make his case, which he presented to Groves, not Oppenheimer. In a telex, he argued that the designers of the gauges should be on board in case anything went wrong: "We couldn't expect these very junior enlisted men to do as good a job as we would." After Groves agreed, Alvarez found himself near the center of the story, thanks to a relationship that he had cultivated outside the official chain of command.

The lieutenant colonel who would fly the bomb, Paul Tibbets, had served under one of Alvarez's heroes, the legendary aviator Jimmy Doolittle. A few days before they were set to leave, Alvarez asked Tibbets how he planned to "get the hell out of there" ahead of the blast. "He told me that they had not thought about this in any detail, so I said I would look into the matter." After an evening with his drafting kit and trigonometry tables, he advised banking sharply while turning 150 degrees, which would point them directly away from ground zero.

"Paul Tibbets and his commanders adopted my plan," Alvarez claimed, which was an overstatement. In fact, Tibbets had thought intensely about the turn, which his crew spent months rehearsing in hundreds of practice runs over the Salton Sea. Separately, Tibbets showed a sketch of the maneuver to Serber, indicating that he was only asking for comments on a decision that he had already made.

After unfavorable weather caused the mission to be postponed twice, the indications all lined up on August 5. That morning, Alvarez watched from the ground as *The Great Artiste*, the B-29 observation plane, did a test drop over the ocean. Instead of detonating, the dummy bomb was supposed to release a puff of smoke at 1,850 feet, aided by four altimeters, "each costing as much as a new Cadillac." Through his binoculars, Alvarez saw it plummet straight into the sea. Nothing happened—the fuses had failed. "Great, just great," he said to the others. "Tomorrow we're going to drop one of these on Japan, and we still haven't got the thing right."

Later that afternoon, Little Boy was loaded onto the *Enola Gay*. A crayon was passed around for bystanders to scrawl something on the bomb, usually "a rude message to the emperor," although Alvarez contented himself with signing his name. The team ran one last check of the gauges, calibrating the microphones in the vacuum of a bell jar before packing the canisters. A midnight briefing led by Tibbets covered weather information and communications protocols, closing with a prayer by the chaplain. Their target would be the city of Hiroshima.

Alvarez, Johnston, and Agnew picked up their gear—including a parachute, personal life raft, flak helmet and suit, and survival vest with fishhooks and medical supplies—and were driven in a truck to *The Great Artiste*. A third plane, later known as *Necessary Evil*, would carry Bernard Waldman, who was equipped with a stabilized Fastax camera capable of shooting thousands of frames per second. Oppenheimer wanted images of the fireball for analysis, but for reasons that would remain unclear, no usable footage was ever captured.

At the airfield, the bombers were surrounded by lights and photographers, like "the opening of a gas station in Hollywood," which Alvarez worried would alert the Japanese holdouts on the island. Before boarding, he handed James Nolan a letter to give to his family in the event of his death. Once their equipment was stowed, he felt exhausted. Deciding that the most useful thing to do was rest, he found some cushions, climbed into the tunnel above the bomb bay that connected the front and rear of the plane, and dozed off. He was asleep when they took off at 2:47 on the morning of August 6, 1945, two minutes after the *Enola Gay*.

Several hours later, after they rendezvoused with the other planes above Iwo Jima, Johnston woke Alvarez to ask him about calibrating the cathode ray tubes. Alvarez was irrationally angry at being disturbed for "an unimportant question," but after emerging from the tunnel, he saw that the sun had risen, which meant that they had to get to work. As the observation team checked the receivers and oscilloscopes, the mood was strangely calm. They were flying over the ocean at a low altitude, so Agnew pushed the receiving antenna out through the floor, which would be impossible once the bomber was pressurized.

Going to the front of the plane, Alvarez spent the next couple of hours seated next to the pilot, Charles Sweeney, until they neared Shikoku, the smallest of the main islands. He returned to the rear, where they spread flak mattresses over the floor as shielding from antiaircraft fire. They also donned protective suits and helmets, although Alvarez elected to skip the parachute—if they were shot down, he didn't want to be taken alive. The windows in the compartment were taped over, so they wouldn't need the dark glasses issued to the rest of the crew.

As they crossed the inland sea to Honshu, closing rapidly on Hiroshima, Alvarez listened through his headphones to updates from Tibbets, who was flying a mile ahead. He heard the signal tone, set to last for sixty seconds, that marked the beginning of the automatic release sequence. The bombardier of the observation plane, Kermit Beahan, opened the bomb bay doors. When the tone ended at 8:15, indicating that Little Boy was falling, Beahan dropped the parachute gauges.

Alvarez started a stopwatch. The bomb would fall for thirty thousand feet, giving them about forty-five seconds until detonation. As arranged beforehand, the planes made a diving turn, the *Enola Gay* to the right, *The Great Artiste* to the left, banking at sixty degrees. Accelerating, the crew felt twice the force of gravity, making it hard to work their equipment until the plane had straightened out again. As they headed away from ground zero, the observation team tuned the receivers to compensate for the fall rate of the gauges. On the oscilloscopes, they saw a rectangular calibration pulse, produced inside each gauge by a burst of air from a piston.

Fifteen seconds later, a flash of white light passed down the tunnel from the cockpit. At that moment, Alvarez knew that the bomb had successfully detonated, but for now, the entire world was reduced to the image on his oscilloscope screen. He saw an N-shaped pulse, rising sharply with the shock wave and falling into negative territory from the area of low pressure that followed, and then a second bump caused by the reflection at ground level. It meant the bomb had gone off in the air, as von Neumann had recommended, to maximize the range of destruction.

Alvarez braced for what he knew would come next. A few seconds after the first shock appeared on the scope, it hit the plane itself, which bounded violently upward from the impact of the spherical wavefront. Even at a distance of eleven miles, it was like sitting on a trash can that was smacked by a baseball bat—the whole plane seemed to crinkle, making a cracking sound like the snap of sheet metal. Two seconds later, it was jolted again by the shock wave reflected from the ground.

After taking a second calibration signal, Alvarez secured his equipment and crawled to the front of the bomber, which had circled to face Hiroshima. The mushroom cloud—deep black, tinted with purple—had reached their altitude, but when he looked down, he saw only an empty wilderness. His mind turned at once to Lawrence's painstakingly separated uranium: "I thought the bombardier had missed the city by miles—had dumped Ernest's precious bomb out in the empty countryside—and I wondered how we would ever explain such a failure to him."

Sweeney quickly set him straight. The bomb had been right on target, and Hiroshima had been completely destroyed. Beahan agreed: "It was a beautiful job of bombing." As they flew once around the cloud, circling the devastated city, the tail gunner, Albert Dehart, filmed it with Agnew's 8 mm camera, while Johnston took snapshots from a side window. The images, which were unauthorized, would be kept secret by the team until they developed them at Los Alamos.

They finally headed south for Tinian. Once they were over the water, the crew members removed their flak jackets for the flight home, deliberately holding back so that the *Enola Gay* would land first. In the rear compartment, Johnston wrote to his daughter, telling her that they had just used an atomic bomb to destroy a city: "We hope that it will not be necessary to drop another one, and that this may be the coup de grace that ends the war."

Taking a sheet of tablet paper, Alvarez started to write a similar letter for Walt. It began:

> This is the first grown-up letter I have ever written to you, and it is really for you to read when you are older. During the last few hours I have been thinking of you and your mother and our little sister Jean. It was tough to take off on this flight, not knowing whether I would ever see any of you again. But lots of other fathers have been in the same spot many times before in this war, and I had a job to do, so I can't claim to be any sort of hero.

Alvarez continued with an account of the mission and some reflections on its meaning, which were still taking shape in his own mind. He wrote until they passed over Iwo Jima, and then he set the letter aside.

Any larger considerations were swept away on their return to Tinian. They landed at four in the afternoon—twelve hours after their departure—to find a party in progress for the 509th. There were hot dogs, sandwiches, pies, and four beers for each man. Almost no one there knew the reason for the celebration, and Alvarez claimed that he was never even debriefed.

After standing in line for food and coffee, they returned to work. Anxious to see the results, they developed the oscilloscope films in the darkroom, finding that only one was good enough to measure the pressure wave. They gave the numbers to William Penney, the head of the British delegation to the Manhattan Project, who was on the island as an observer. Penney eventually came up with an energy equivalent to 8,700 tons of TNT. Once again, it was about forty percent below the best later estimates, which was as good as the gauges ever managed.

Sixteen hours after the bombing, President Truman announced it to the world, stating without attribution that it had "more power than 20,000 tons of TNT." The base went wild, but Project Alberta was already turning to the next bomb. To demonstrate that the attacks could continue indefinitely, Truman's advisors wanted to drop two atomic weapons in succession, which was why they had waited until the uranium was available for Little Boy. Now it was time for Fat Man, the plutonium bomb, which utilized Alvarez's exploding bridgewires.

An ordinary man might have been ready to step back, but Alvarez wanted to be part of the story one last time. On August 8, the night before the second bombing, he was drinking a beer at the officers' club with Serber and the physicist Philip Morrison. He suddenly remembered Ryōkichi Sagane, the scientist he had met in Berkeley in 1936. Sagane was currently a professor in Tokyo, and Alvarez thought that it might be possible to send him a message.

The planes were scheduled to depart in two hours. After obtaining permission from an intelligence officer, Alvarez retrieved a pen and paper from his tent. With contributions from Serber and Morrison, he addressed a letter to Sagane, along with two carbon copies, identifying its source as "Headquarters, Atomic Bomb Command." Referring to themselves as "three of your former scientific colleagues," they urged Sagane "to convince the Japanese General Staff of the terrible consequences which will be suffered by your people if you continue in this war." The letter continued:

> You have known for several years that an atomic bomb could be built if a nation were willing to pay the enormous cost of preparing the necessary material. Now that you have seen that we have constructed the production plants, there can be no doubt in your mind that all the output of these factories, working 24 hours a day, will be exploded on your homeland.
>
> Within the space of three weeks, we have proof-fired one bomb in the American desert, exploded one in Hiroshima, and fired the third this morning.
>
> We implore you to confirm these facts to your leaders, and to do your utmost to stop the destruction and waste of life which can only result in the total annihilation of all your cities, if continued. As scientists, we deplore the use to which a beautiful discovery has been put, but we can assure you that unless Japan surrenders at once, this rain of atomic bombs will increase manyfold in fury.

They left the message unsigned. Inserting the three copies into envelopes, they secured them with several layers of Scotch type on the aluminum canisters of the parachute gauges. Alvarez made sure that they were loaded, untouched, on *The Great Artiste*, which would fly for the second time as an observation plane, with Johnston on board to operate the equipment.

At the briefing, the crew was told that the primary target would be the castle town of Kokura. There were rumors that the Japanese would attack Tinian, where a blackout was ordered to increase security. Using a flashlight, Alvarez helped Serber—who was scheduled to ride on a third plane with a Fastax camera—retrieve and stow his gear. Alvarez and Agnew then drove off to watch the bombers depart.

They took their truck to a hill south of the airfield, which offered an excellent view. When they were partway up the slope, Alvarez turned to face downhill, switching off his headlights. In the silence, he thought that he heard movement in the sugarcane. Picturing the Japanese soldiers who still lurked on Tinian, he quietly released the hand brake and coasted farther down the road. Although he still seemed to hear voices, he watched the planes take off without incident. He later concluded that it had all been his imagination: "I have never been so frightened in my whole life."

When Alvarez and Agnew returned to camp, they were surprised to find Serber there. He explained that he had forgotten his parachute—he had somehow picked up a life raft instead—and was ordered off the plane by the pilot. The engine noise drowned out his protests, so he was unceremoniously tossed onto the runway. No one else on board knew how to operate the camera, which was the only reason that the third plane was there in the first place.

It was an inauspicious beginning for what Alvarez called "as abominably run a raid as any in the history of strategic warfare." A series of mistakes led them to abandon Kokura for the secondary target of Nagasaki. Since it was obscured by cloud cover, they decided on a radar drop, in violation of their orders to conduct a visual bombing. Otherwise, unthinkably, they would have had to dump the bomb in the sea. At the last second, a gap in the clouds reportedly allowed Kermit Beahan to use his bombsight. Alvarez was doubtful, noting that they were off by two miles—a reasonable error for radar, but not "a visual drop by one of the best bombardiers in the Air Force."

As Fat Man plunged toward Nagasaki, *The Great Artiste* released the three parachute gauges. The microphones all failed—the signals from the explosion were off the scale—but the canisters were later discovered and taken for examination to a naval base, where the letters were opened. Although they were allegedly read by the high command, Sagane wasn't informed for a month and a half, and they had no apparent impact on the debate in Japan.

The plan was to keep dropping atomic bombs as more plutonium cores arrived, but it was all put on hold after the Japanese announced their unconditional surrender on August 15. As soon as he heard Emperor Hirohito's speech, Alvarez sent a telex to Groves, asking for permission to leave. In response, he was ordered to stay until the war was officially over. Until the formal surrender ceremony, Groves wanted the technical crew on Tinian, where they would serve as a warning to the Japanese of the consequences of changing their minds.

Alvarez did what he could to pass the time, swimming at the beach during the day and playing poker at night. He also reached an arrangement with the Japanese prisoners of war to trade soap for cowrie shells, six of which he made into an ornament for Geraldine. After the instrument of surrender was signed at the beginning of September, he was cleared to leave at last.

In Hawaii, he received a hero's welcome, but he was troubled by the mood at Los Alamos. The scientists there seemed "almost neurotic" over the bombings, leading him to take an even harder line in their defense. "Many of my friends felt responsible for killing Japanese civilians, and it upset them terribly. I could muster very little sympathy for their point of view; few of them had any direct experience with the war and the people who had to fight it." Finding that he could no longer talk to these colleagues, he avoided them as he arranged to return to Berkeley. He could finally tell Geraldine about his work, but their marriage had been hurt by years of secrecy.

While he hoped on some level to pick up where he had left off, the bomb had changed him forever. His feelings about its morality resembled the conclusion reached by Oppenheimer, Lawrence, Compton, and Fermi—the four scientists who had played the greatest role in his career—in an official recommendation in June: "We can propose no technical demonstration likely to bring an end to the war; we see no acceptable alternative to direct military use."

Alvarez also felt that two bombings were necessary. "If I had been a Japanese nuclear physicist advising the war ministry, I would have told them how expensive and difficult it was to separate uranium isotopes—plutonium was still an Allied secret—and would have estimated that another such bomb as Little Boy probably wouldn't be ready for months." Hiroshima wouldn't have been enough without the confirmation of Nagasaki, which Alvarez underlined by emphasizing "the production plants"—not the bomb itself—in his letter to Sagane.

To explain his convictions, Alvarez drew an analogy with the San Francisco earthquake of 1906, which ignited a citywide inferno. "The fire started because the earthquake damaged gas lines; it grew into a major conflagration because the earthquake also damaged water mains to such an extent that firemen couldn't use water to fight the fire. With no other recourse at hand, they began dynamiting houses in the fire's path to make firebreaks."

Alvarez concluded that the atomic bombs served as a similar bulwark against the plan of war that the United States was actively pursuing. "The hundreds of B-29s on the four airfields in the Marianas correspond to the approaching fire, which would have destroyed far more lives in Japan in the weeks to come by continuing to fly incendiary missions than were sacrificed in the creation of a firebreak." For Alvarez, the conflagration wasn't a hypothetical scenario, but the inevitable outcome of the machine that his country had set in motion.

At Tinian, he had watched the planes depart in endless waves. The firebombing had already killed hundreds of thousands, including 90,000 on a single night in Tokyo. He had every reason to believe that if the bombs hadn't been used, Hiroshima and Nagasaki would have been burned immediately, with the Allies prepared to kill far more than the 150,000 to 220,000 who died in the atomic bombings. By that standard, Alvarez contended, Truman's decision had saved lives.

The real question was what the future would bring. Alvarez had tried to articulate his thoughts in his letter to Walt, and although his views would evolve over time, they would never truly change. He saw the bomb as the ultimate solution, a weapon that would make war unthinkable by scaling up the cost to an unimaginable extent. It was a dream that had been frustrated before, and by the time his son was old enough to read his prediction, Alvarez—and the rest of the world—would have a better sense of whether he had been right or wrong:

What regrets I have about being a party to killing and maiming thousands of Japanese civilians this morning are tempered with the hope that this terrible weapon we have created may bring the countries of the world together and prevent further wars. Alfred Nobel thought that his invention of high explosives would have that effect, by making wars too terrible, but unfortunately it had just the opposite reaction. Our new destructive force is so many thousands of times worse that it may realize Nobel's dream.

5.

STAR CHAMBER

1945–1954

REFLECTING MUCH LATER ON HIS TRIUMPHANT RETURN TO Berkeley, Alvarez wrote of himself and his colleagues, "We had gone away as boys, so to speak, and came back as men. We had initiated large technical projects and carried them to completion as directors of teams of scientists and technicians. We were prepared to reassume our subordinate roles with Ernest as our leader once again, but by his actions, if not in so many words, he signaled that we were to be free agents." Liberated from the urgent problems of the war, Alvarez was suddenly guided by nothing but his own taste.

As he took a streetcar to campus from his new house on Cedar Street, he felt secure in his position. Earlier in 1945, Lawrence had been asked by James B. Conant, the president of Harvard, for recommendations for a university appointment. Out of honesty, Lawrence put Oppenheimer and Alvarez at the top, while noting that "it would be no less than a calamity if either Oppie or Alvarez should not return here." To eliminate any risk of his departure, Alvarez was made a full professor.

For the rest of his life, he would be known as one of the men who built the bomb, and he had no intention of being excluded from the postwar debate. In 1946, he was appointed by Groves to the Technical Committee on Inspection and Control, which would evaluate proposals for regulating atomic research. The United States currently held a monopoly, but it would not go unchallenged for long.

At the president's request, Oppenheimer and Secretary of State Dean Acheson consulted Alvarez's group in New York on the challenges of monitoring what was "an essentially living art." In its report, the committee, which Acheson chaired alongside David E. Lilienthal, recommended sharing information on the bomb with the Soviet Union, with both countries holding off on further development. It was too radical to be adopted, but it reflected the rapidly evolving situation, as scientists reconsidered the open exchange of ideas that they had once taken for granted.

In August, Berkeley hosted a delegation of Soviet researchers. On a tour, the physicist Mikhail Meshcheryakov saw a prototype of the magnetic spectrometer used to produce uranium-235. Looking at it carefully, he asked why the cyclotron magnet was horizontal, not vertical—a feature that allowed multiple separation tanks to be put in a row. Thinking fast, Alvarez replied that they were looking into a new way of mounting the heavy electrodes. "We both knew that I was lying," Alvarez recalled, "even though it was a plausible story."

Alvarez was looking to benefit from his elevated profile. Although he turned down a job at General Electric, he pitched the company on using radar stations to establish parallel lanes for air traffic, like highways in the sky. His proposal, eventually known as Tricon, was his first attempt to find a commercial outlet for his ingenuity, but it failed to get any traction.

In Berkeley, by contrast, there was no sign of the fall in funding that he had feared. On the premise that science was crucial to national security, they would be generously supported by the new Atomic Energy Commission. Alvarez missed the improvisational mood of the old days, but he was ready for what promised to be a golden age at the lab: "We ran it with a big barrel of greenbacks."

The tradeoffs would soon become clear, but for now, Alvarez knew exactly where to start. "All of us had gone off to war pretty much secure in the knowledge that when we came back we would build big accelerators and we would explore the thing then called the mesotron." This mysterious particle—later known as the muon—had been observed as a negatively curved but surprisingly massive track in cosmic rays. Producing it in a lab would require higher energies than ever.

While Rabi hoped for "some easy way" to generate powerful beams, Alvarez told him that he favored the direct approach: "Until someone invented it, the best thing to do was use brute force." To achieve it, he turned to the linear accelerator, or linac, which Lawrence had once rejected as impractical. Because a cyclotron was round, Alvarez reasoned, its cost rose sharply with the cube of its radius. A linear accelerator, by contrast, could be as long as you wanted, with the expenses rising in a steady line. At some point, it would be more economical; the only question was when.

Alvarez found a way to make it even more attractive. As he had told Lawrence, the Signal Corps owned thousands of obsolete SCR-268 radars. With help from Groves, he eventually procured an entire warehouse of surplus antennas, which Lawrence hailed as a sign of the university's strong relationship with the military. Alvarez planned to use six hundred sets as oscillators for a huge electron linear accelerator, spanning two thousand feet in a canyon in the hills, that would reach the energy threshold for producing mesotrons. Like the cyclotron, it might be useful in unanticipated ways, and he wanted to control it.

All the while, McMillan had been considering the problem from a different direction. Lawrence's powerful new 184-inch cyclotron could still accelerate particles for just a few revolutions. As Einstein had predicted, their effective masses increased as they reached relativistic speeds, causing them to fall out of step with the electric field. The result seemed to be a hard ceiling on beam energy.

McMillan came up with a brilliant solution. The accelerating field could be pictured as a waveform, with the particles receiving a push at the crest of each wave. He realized that if they arrived at the side of the wave instead—where the field was changing—it would lock them into the same phase, as particles received more or less energy depending on when they arrived. With the right adjustments, you could keep accelerating them indefinitely along a single orbit.

"When Ed told me about it," Alvarez wrote, "I shook his hand." The concept of phase stability was "so elegant, inexpensive, and obviously workable" that he didn't think his electron accelerator could compete. A full version of what McMillan called the synchrotron was still years away, but in the meantime, they could put the principle to work in the 184-inch cyclotron, which was converted into a "synchrocyclotron" with aspects of both designs. Alvarez wasn't sure whether this made sense for larger particles, however, so he switched his own plans overnight to protons.

The decision left him in a bind. He had committed to building a linear accelerator using radar sets, which was "perfectly reasonable" for electrons, but not for heavier particles. Pulling it off would require vastly more power. If he hadn't already committed himself, it wouldn't have occurred to him to try, but as with blind landings—which had unexpectedly called for a totally new antenna—he saw no other choice: "Once you tell people you're going to do something, then you had better do it."

Alvarez rapidly assembled a team, including Oppenheimer's brother, Frank, and his most brilliant associate from the war, Pief Panofsky, whom he saw as his "secret weapon." They set up shop in Building 10, a wooden structure in the hills that was seventy feet long, placing an upper limit on the machine's dimensions. After developing a powerful electrostatic generator that could inject protons at high energies, they turned to the accelerator itself.

Although the plan had been to use surplus radar sets, the tubes constantly burned out, so Alvarez dropped the idea entirely—he never even visited his warehouse of antennas. Inside the accelerator, the protons entered a series of copper drift tubes that shielded them from the current, except at strategically placed gaps. The cavity around the tubes was enclosed in a vacuum chamber, reinforced by aluminum rings, that the team compared to "an inside-out airplane."

In the fall of 1946, they took delivery of the cavity, which was supported at either end by electrical hoists. One day, Oppenheimer received a frantic phone call from his brother. Tracking down Alvarez at the physics building, Oppenheimer delivered the bad news. A hoist had locked into the "on" position, rising until a chain broke and an iron support rod crashed down. Rushing to the building, Alvarez found that the cavity had been irreparably crushed.

Unexpectedly, the catastrophe had a positive impact on his private

Staff posing on the forty-foot vacuum tank of the linear accelerator outside the 184-inch cyclotron building in 1946.
Lawrence Berkeley National Laboratory.

life. Years of wartime secrecy had strained his marriage to Geraldine, who missed her mother in Chicago. "In recent weeks she had talked of divorce, which I didn't want but didn't know how to prevent. When I arrived home on the afternoon of the accident in a state of gloom, Geraldine provided great comfort, realizing that she was also needed in Berkeley." Instead of separating, they reconciled, moving into a house on Vassar Avenue that they saw as their permanent home.

On a deeper level, the incident reflected Alvarez's assumptions about Geraldine's role in his life. Jean, their daughter, felt that while Geraldine was close to the wives of other physicists, she had little interest in playing the part that Alvarez expected: "There was a resistance in her to being the perfectly supportive wife." Reflecting on the linear accelerator story, which occurred when she was only two years old, Jean said, "[Geraldine] found space in herself to respond to his emotional needs in 1946, but by the time I was aware, I don't think I saw much of that."

Alvarez's team built a better cavity, but he had still received just half of his funding. On a tour of the lab, he made his case to the AEC commissioners, including Lawrence's close friend Lewis Strauss, a fifty-year-old banker who had emerged as a power broker in atomic research. During a summer conference in 1947 at Bohemian Grove, an exclusive campground in Sonoma County, Alvarez and McMillan gave presentations to the skeptical Strauss, who thought Berkeley was "running out of the duty it owed." Lawrence, undaunted, cut a deal for both projects.

To take advantage of phase stability, Alvarez was discarding basic aspects of his original design. Protons had to cross the gaps as the waves were rising, giving late particles a boost, but he also needed to focus the beam, which was usually accomplished by exposing it to the field as it fell. By definition, it was impossible to do both at once. At first, he tried focusing with beryllium foils, which were fried to a crisp, so his team found a work-around using grids of tungsten strips, with an attitude of "damn the torpedoes, full speed ahead."

In October 1947, they finally turned on the beam, which went only a short distance. Gathering the group around a blackboard, Alvarez announced that they would redesign it all from scratch. Instead, they came up with a few minor fixes that produced a working beam that night. Afterward, a photo of his lecture circulated with a sardonic caption: "9 p.m.—Luie explains why accelerator can't work. First beam obtained at midnight the same evening."

Even after it was running, the linear accelerator presented considerable safety issues. A technician passed out from inhaling the nitrogen gas that insulated the generator, which was stored in pressurized cylinders. One day, two men were walking nearby when a canister blew up, setting off a chain of explosions. No one was hurt, but the wall was demolished.

Using the 32 MeV beam, Alvarez bombarded targets to make boron-8, a source of solar neutrinos, and nitrogen-12, "which had the shortest half-life and highest beta-ray energy ever seen." All the same, he knew that it wasn't "the world beater that we thought it was going to be." When the first scintillation counter—a detector capable of registering very faint pulses—appeared in Berkeley, McMillan joked, "Now we have a device to detect the beam from the linear accelerator."

One day, Lawrence casually said to Alvarez, "By the way, did I tell you that I've given your accelerator to the University of Southern California?" According to Panofsky, the decision was "rather controversial," since reinstalling it would cost as much as it took to build it in the first place, but Lawrence wanted it off his hands. Alvarez felt that it had fallen through the cracks: "The linac was a sort of orphan here." He claimed not to mind, but his pride was wounded. McMillan's synchrotron would change accelerator physics forever, while the linac was barely a footnote.

When Lawrence nominated Alvarez and McMillan for the National Academy of Sciences, he told each man to write the other's recommendation, which only reminded Alvarez of their relative positions. He consoled himself with his public honors. For his work on blind landings, he received the Collier Trophy of the National Aeronautic Association from President Truman in the Oval Office. He was also awarded the Medal for Merit, the highest decoration for civilians in the war effort, for his contributions to the MIT Rad Lab, which seemed like a distant memory.

"Science is like baseball," Alvarez later said. "If you haven't made it and become a manager by the time you're thirty-five, you might just as well quit and run a garage or a filling station." As he passed that age, his need for recognition was greater than ever. The author Nuel Pharr Davis, who rarely had kind words for Alvarez, later wrote that he "divided his talent between experiments and public relations," and this side of his personality became apparent even to his admirers.

One was the astronomer Gordon Pettengill, a doctoral candidate at Berkeley, who remembered how Alvarez always arranged to be paged thirty minutes into the weekly physics meeting—a trick that he had learned from the station commander at Elsham Wolds. "This served two purposes," Pettengill recalled. "One, it put his name before everyone so they knew he was there, and secondly, it gave him a chance to leave. So he was credited with being, you know, how should I say it, a little more sophisticated than many of the other scientists."

Like Lawrence, who collaborated with Alvarez on a study of radiological warfare, he was moving in high circles. At a summer conference in Zurich in 1948, he encountered Werner Heisenberg, one of the leaders of the German nuclear project, who was privately derided for claiming to have deliberately held back the Nazi bomb. "I didn't look forward to meeting him, even though I still admired him for his invention of quantum mechanics. We had a fruitful exchange of ideas, however, avoiding the points that would make our interaction unpleasant."

Before joining Geraldine for a sightseeing tour of Europe, Alvarez headed for England, where he dined with Bill Penney, the head of the British atomic program. They had served together on Tinian, but their countries had ceased to share nuclear information after the war. When Penney said that he didn't know what kind of ceramic to use in crucibles for plutonium, Alvarez kept the answer to himself, although it could have been found "in five minutes in the Los Alamos library."

As the national mood was darkened by suspicion toward the Soviet Union, more overt displays of loyalty were required. In March 1949, Robert Sproul, the university president, was warned that the state legislature was targeting communist activity in the educational system. He decided to preemptively revise the oath taken by faculty members, inserting a clause disavowing "any party or organization that believes in, advocates, or teaches the overthrow of the United States Government."

Sproul assigned this show of patriotism to John Neylan, the chairman of the board of regents, who added language that explicitly rejected the Communist Party. Anyone who refused to sign would be dismissed, leading to a firestorm over academic freedom that raged for two years. As Alvarez wrote, "Important faculty members resigned, attempts to recruit attractive young faculty often failed, and many of my colleagues turned on each other with undisguised hatred."

Alvarez signed it without any qualms, while Lawrence forcefully supported Neylan, his longtime mentor. One day, Lawrence asked Gian Carlo Wick, an Italian physicist in Berkeley, if he had signed the oath. When Wick said that he hadn't, Lawrence responded coldly, "Then you can no longer work at the Radiation Lab." On hearing about the exchange, Alvarez warned Lawrence that he had overstepped his authority. "Ernest grunted and groaned and made a lot of noises for a while and eventually calmed down and agreed that I was right."

Both Oppenheimers were already gone—Robert to the Institute for Advanced Study in Princeton, Frank to Minnesota—but neither would escape the shift in the political climate. The first to feel it was Frank, who testified to the House Un-American Activities Committee that he had once been a member of the Communist Party. After he lost his faculty position, he learned that Lawrence had unilaterally ruled out his return to Berkeley. He also heard that Alvarez had passed along a rumor—from an unidentified "security person" on a train—"that Frank had been a spy."

Other scientists in Berkeley resigned or were fired. One who refused to sign the oath was Jack Steinberger, a future Nobel laureate, who recalled being "lectured by Alvarez on the evils of the fifth column of communist 'sympathizers,' in whose traitorous ranks he probably included me." After Alvarez banned him from the cyclotron, Steinberger left for Columbia. According to Steinberger, one observer felt that Alvarez—who admitted that they lacked "good chemistry"—was motivated by personal animosity: "Serber wrote me that it was jealousy."

Reflecting afterward on the debacle, Segrè said, "For theory, it was a body blow; for experiment, something a bit less." And it occurred at the worst possible moment. On August 29, 1949, the Soviet Union detonated a plutonium bomb, based largely on stolen information about Fat Man. Radioactive debris from the explosion, which was equivalent to twenty-two thousand tons of TNT, was detected by an American reconnaissance plane—a descendant of the airborne detection system that Alvarez had designed for Groves. When confirmed, it sent shock waves through the intelligence community, which had assumed that a Soviet bomb was years away.

Despite his government connections, Alvarez heard the news only after Truman made the official announcement, while Lawrence learned about it at a roadside newsstand on a trip to Yosemite. The prospect of a peaceful understanding with the Soviets seemed very far away, and in the absence of better answers, the two men seized on an approach that they had used before. When all else failed, they had often succeeded by scaling up an existing solution, so perhaps it was inevitable that their thoughts turned immediately to the thermonuclear bomb.

ALTHOUGH THE HYDROGEN BOMB would be forever associated with Edward Teller, its origins went back to an offhand remark by Fermi. In September 1941, the two men were working together on the pile at Columbia University. One afternoon, while walking to the lab after lunch at the faculty club, Fermi suddenly spoke up. "Now that we have a good prospect of developing an atomic bomb, couldn't such an explosion be used to start something similar to the reactions in the sun?"

Teller understood at once. In the intense pressure of the sun's core, protons fused into helium nuclei, producing light and heat. Fermi was wondering whether a fission bomb could ignite a mass of hydrogen, starting a fusion reaction that would be theoretically unlimited in its destructive power. Teller was galvanized by the idea of the "Super," which he pursued all but single-handedly for the rest of the decade. By the time of the first Soviet nuclear test, as Alvarez later observed, "the program essentially did not exist except for Teller."

On October 5, 1949, Alvarez lunched with Wendell Latimer, the dean of chemistry, a dedicated anticommunist who had overseen wartime research into plutonium at Berkeley. They discussed the possibility that the Russians would follow up on their fission weapon by proceeding to the hydrogen bomb. That night, Alvarez began a diary, believing that it would provide a useful historical record, as well as documentation for any patentable inventions. "The only thing to do," he wrote, "seems [to be] to get there first—but hope that it will turn out to be impossible."

At that point, he was thinking largely in terms of alerting his contacts. The previous month, he had received a call from Carl Hinshaw, a member of the U.S. House of Representatives from California, to ask for his thoughts on air navigation. Alvarez was preparing an illustrated report for Hinshaw, who also served on the Joint Committee on Atomic Energy. Lawrence—a master at informally working the levers of power—agreed to help Alvarez advocate for the "neglected" hydrogen bomb on an upcoming trip to discuss radiological weapons in Washington.

They called Teller at Los Alamos, but the phone connection was bad, so they left early to meet him in person. That night, Alvarez—embarking on yet another project that he couldn't reveal to Geraldine—flew with Lawrence to New Mexico. The next morning, they saw Teller, George Gamow, and Stanisław Ulam. As they listened intently, Teller updated them on the thermonuclear program, for which he hoped to utilize MANIAC, a computer based on a design by von Neumann.

The conversation with Teller continued at a hotel in Albuquerque, where Lawrence took off his shirt to wash it in the bathroom. Pointing out that Teller would need to be on the road constantly to promote the hydrogen bomb, Lawrence advised him to buy shirts made from drip-dry fabric: "I don't need to carry half a dozen shirts along, but can wash them every night as I did just now." At that moment, Teller remembered, he realized that they were serious.

As a starting point, Teller recommended building reactors moderated by heavy water, which would be more effective than graphite at producing neutrons for making tritium, the hydrogen isotope that fueled the fusion reaction. Lawrence and Alvarez decided to design heavy water reactors at Berkeley, establishing control over a piece of technology that might be useful in other ways, as well as energizing the physics department at a difficult time.

On October 8, Lawrence and Alvarez flew to Washington, where they met with AEC officials and dined with Alfred Loomis. It seemed best to operate outside the usual channels, and Lawrence—whom everyone saw as the man in charge—naturally took the lead. While neither he nor Alvarez was a reactor expert, it would be simple enough to copy an existing design. By consensus, the best heavy water pile was at Chalk River, outside Ottawa, which they hoped to visit.

Given the diplomatic issues involved, they needed supporters in high places. Alvarez phoned Congressman Hinshaw, who met them for lunch with Senator Brien McMahon of Connecticut—the author of the Atomic Energy Act of 1946—and his protégé William Borden, the hawkish staff director of the Joint Committee on Atomic Energy. According to Borden's notes, the visitors were concerned that the Soviets might already have a head start on the hydrogen bomb: "They declared that for the first time in their experience they are actually fearful of America's losing a war."

A meeting with David E. Lilienthal, the AEC chairman, was less encouraging. Alvarez was shocked by his reaction: "He turned his chair around and looked out the window and indicated that he did not want to even discuss the matter." Privately, Lilienthal saw them as "bloodthirsty," writing in his diary, "Ernest Lawrence and Luis Alvarez in here drooling over the [hydrogen bomb]. Is this all we have to offer?" The other commissioners were more receptive, including Lewis Strauss, who had already advocated "a quantum jump" in the thermonuclear program.

Heading for New York, they planned to fly to Ottawa, but were unable to get tickets in time. Instead, they met with I. I. Rabi, who struck Alvarez as "very happy" about their return to atomic research: "It is certainly good to see the first team back in." The encounter left a slightly different impression on Rabi, who later said, "I generally find myself when I talk with these two gentlemen in a very uncomfortable position. I like to be an enthusiast. I love it. But those fellows are so enthusiastic that I have to be a conservative."

Alvarez flew back that night to Berkeley, where he found McMillan, Serber, and Seaborg willing to participate in a crash program to build a reactor. Instead of trying to make the perfect pile, they would assemble it in units that could be rapidly reconfigured: "The one thing we could supply was the ability to build large-scale apparatus and build it fast." On October 18, Lawrence formally put Alvarez in charge. In his diary, he wrote, "I am therefore going on almost full time as director of a nonexistent laboratory on an unauthorized program."

It felt like a return to the heady atmosphere of the war. Cleaning out his desk, Alvarez moved into the director's office in a new building in the hills. He visited a possible site near Suisun Bay, an hour from Berkeley, that was safely removed from any houses, with plenty of water for cooling. At night, he pored over papers on reactor design. "I didn't want to build reactors. I disliked the idea of building reactors. But I felt the country needed them."

None of this would matter without the support of the General Advisory Committee, a panel chaired by Oppenheimer that provided the AEC commissioners with counsel from scientists, including Rabi, Seaborg, and Fermi. When Teller said that "Oppie was lukewarm" on the proposal, Alvarez was understandably alarmed. Calling Oppenheimer, he asked if they could discuss it before the next committee meeting. "As I remember it, Dr. Oppenheimer said he would be very glad to see me in Princeton, and in fact invited me to stay overnight in their guestroom."

Shortly afterward, Alvarez realized that his schedule was too tight for a visit—he was planning to meet reactor experts in Chicago, which seemed like a better use of his time. In his place, he "deputized" Serber to present the case to Oppenheimer, reasoning that the two men were good friends. Serber recalled it somewhat differently, writing that Lawrence dispatched him because "he thought I would get a more sympathetic hearing from Oppenheimer than Luis would."

After two days with the Chicago group, Alvarez flew to Washington on October 28. At AEC headquarters on Nineteenth and Constitution Avenue, he and Serber delivered presentations to the scientific advisors. When Fermi asked why a heavy water pile should be built at a university with no track record in reactors, Serber said that they only wanted to move fast: "If there were a better and quicker way of accomplishing that end, Ernest would be the first to be in favor of it." Privately, he knew that the question implied that the Berkeley program was probably a nonstarter.

The next day, as the committee began to deliberate, Alvarez and Serber waited in the lobby. At the main entrance, Alvarez spoke briefly with Borden, who was there on behalf of Senator McMahon, and kept an eye on the arrivals: "I watched my friends go upstairs, and I saw the famous military men"—the Joint Chiefs of Staff—"whom I recognized from their pictures follow along." Hours later, when several members emerged from the elevator, Alvarez asked brusquely, "Yes or no?" Instead of answering, Oppenheimer invited them to have lunch nearby.

In the small, dim café, Alvarez listened in disbelief as Oppenheimer expressed strong reservations. "If we built a hydrogen bomb [Oppenheimer said], then the Russians would build a hydrogen bomb, whereas if we did not build a hydrogen bomb, then the Russians would not build a hydrogen bomb. I found this such an odd point of view that I don't understand it to this day." Given Oppenheimer's support for the idea during the war, Alvarez wondered if he felt guilty over Hiroshima and Nagasaki, and he was astounded when Serber agreed that the program was inadvisable.

Although the discussions would continue for another three days, Alvarez left at once, sensing that the proposal was doomed. In his last diary entry, he alluded to a "particularly interesting talk with Oppie," describing it as "pretty foggy thinking." Oppenheimer later said that he had only been relating "what other people were saying," but acknowledged that he had been "indiscreet." He expressed his real concerns to James B. Conant: "What does worry me is that this thing appears to have caught the imagination, both of congressional and of military people, as *the answer* to the problem posed by the Russian advance."

In its confidential report to the commissioners, the GAC came out decisively against a thermonuclear device, arguing that it had no justifiable military use. If it was developed, it concluded, America would undermine its moral standing by pursuing "a weapon of genocide." For Alvarez, it was a wake-up call. He had overestimated his influence, not realizing that he was still viewed—unlike Oppenheimer—as a marginal player outside Berkeley.

As Alvarez moved his files back to his old office, he wondered about

Oppenheimer's change of heart. A month later, he and Lawrence saw Vannevar Bush at Stanford. On a drive back to his hotel, Bush mentioned that he had chaired the committee that verified the Soviet bomb test. According to Alvarez, Bush said, "I think the reason the president chose me is that he does not trust Dr. Oppenheimer."

Bush later denied this statement, noting that the committee had been assembled not by Truman but by General Hoyt Vandenburg, who had included Oppenheimer as well. Alvarez conceded that he might have misremembered the name of the authority in question, but not the point of the remark, which Lawrence also recalled. "This was the first time that anyone had ever said in my presence," Alvarez stated, "that Dr. Oppenheimer was not to be trusted."

The truth, of course, was that only one man's opinion really mattered. In the end, Truman ignored the GAC's recommendation and approved the thermonuclear program. The drama in October had minimal impact on the hydrogen bomb—it didn't even slow it down—but it was hugely meaningful to the men involved. Alvarez was sidelined, and he had good reason to see Oppenheimer as responsible.

Although Lawrence had lost, he badly wanted to stay in the game. He found another cause in a looming shortfall in uranium, which depended on unreliable foreign mines. As a solution, he proposed building a neutron source to make fissionable materials, including tritium. What he really wanted was a project that yielded as many neutrons as a reactor but could be justified for his own lab.

Returning to his trick of scaling up an existing approach, Lawrence, who had repurposed the cyclotron to produce uranium isotopes during the war, envisioned doing much the same with a gargantuan version of Alvarez's linear accelerator. By bombarding heavy-element targets with deuterons, it could generate plenty of neutrons. Emphasizing that money was no object, Lawrence wondered aloud if they could make the lining out of gold. According to his biographer, Herbert Childs, his associates "frankly stared, openmouthed."

Instead of designing a heavy water pile, Alvarez would run this even less likely project, euphemistically known as the Materials Testing

Accelerator. After Oppenheimer wrote a "negative but not damning" evaluation, Lawrence extracted $10 million from the AEC for the Mark I prototype, which was only a preview of the machine that he hoped to build in Missouri. In his imagination, the Mark II would be a quarter of a mile long and consume as much power as a small city.

As Childs later wrote, Lawrence was staking everything on an approach that had succeeded magnificently elsewhere: "Forget obstacles, build a basic part of what is desired, and proceed, inventing around difficulties as they arise." Partnering with Standard Oil of California, Lawrence and Alvarez visited potential locations in the summer of 1950. One day, they drove forty miles southeast to a former naval air station in Livermore with barracks, a gym, and a disused swimming pool. Looking around the future home of the Lawrence Livermore National Laboratory, the man who would be its namesake declared, "Well, Luie, this is it."

Guided by the invaluable Panofsky, the group vastly enlarged the dimensions of the linear accelerator. Because the energies involved would destroy the tungsten grids used to focus the beam, they pivoted to focusing with magnetic fields, which increased the diameter from two inches to three feet. Under a huge scaffold, the largest vacuum tank in history went up in the open air in Livermore, with the giant tubes moved into place on copper railroad tracks. At sixty feet high and almost ninety feet long, the MTA made the men around it look and feel like ants.

Operating it was even more humbling. Whenever an accelerator was switched on for the first time, irregularities in the drift tubes generated sparks. The discharges in the MTA were the size of lightning bolts, producing what Panofsky called "spectacular 'stalagmites and stalactites' of copper." To eliminate them, Alvarez cranked up the power at the control table until he heard the "fearsome thunder," then climbed into the tubes to clean them out. He was soon working eighteen hours a day, sleeping on a cot in the barracks, and saw his family less than ever.

Overcoming faulty welds and vacuum leaks, Alvarez gradually improved the setup to run for a full hour. When it produced its first deuteron beam, however, it had already outlived its rationale. Thanks to technical advances and government incentives for prospectors, the uranium shortage disappeared overnight. The MTA—which cost twice as much as originally projected—had been a monstrous response to a problem that was ultimately solved by the free market.

In the end, the project, which wasn't even useful for physics, was canceled entirely. Alvarez wrote that it had "attempted to carry a technology beyond its reasonable limits," which understated his frustrations during what he recalled as "not a happy time." He had taken a hard line on defense work, telling colleagues, "Anyone who now takes the time to work on mesons is little less than a traitor." As he belatedly returned to research, it was tempting to blame Oppenheimer for the "two years I wasted in unsuccessfully attempting to solve the fictious uranium shortage problem."

Alvarez had also been estranged from many of his colleagues. Panofsky believed that the physics department had been permanently damaged by the loyalty oath controversy, and he was unconvinced by the case for the thermonuclear bomb. One night, Panofsky told his wife, "Today is a red-letter day because Alvarez didn't call me a traitor." Eventually, he headed for Palo Alto, unmoved by Alvarez's reaction: "Nobody ever does anything at Stanford, you could kill yourself."

Segrè left as well, partly over his irritation with Alvarez's "high-handed position" that "all the time of the department should be spent on war work." Even Teller felt uncomfortable enough about the loyalty oath to decline a position at UCLA, leading to a tense meeting with Lawrence that Alvarez made a point of attending. "He remained silent while Ernest gave his opinion, and while I gave mine," Teller said. "But Luis's presence helped me temper my heat. I was deeply grateful to him, because, when I left, it was clear that Lawrence would continue his intensive support of the Super."

Another casualty of the troubled atmosphere was Serber, who left for Columbia, feeling caught in a growing conflict between Lawrence and Oppenheimer. At Berkeley, funding was entwined with defense, while the more independent Oppenheimer could afford to take moralistic stands. As Alvarez put it, "Ernest felt that Robert enjoyed being a little tin god, and felt that Robert ought to stick to his knitting instead of going around and giving speeches on philosophy and politics."

Alvarez found himself ensnared in the political side as well. In December 1950, a few months after the outbreak of the Korean War, Oppenheimer had called with an invitation: "We are having a meeting of a committee to try to find out the future of the military applications of atomic energy. I would like to have you on this committee because I know you represent a point different from mine." Alvarez was pleased by the request, which he saw as a conciliatory gesture.

In Washington, Alvarez learned that the Long Range Planning Committee favored "tactical" nuclear weapons over the hydrogen bomb, evidently out of concern that the two programs were competing for resources. Alvarez disagreed—one was a technical problem, the other a matter of fundamental research—and he was alarmed by Oppenheimer's attitude. Instead of canceling the thermonuclear project, Oppenheimer said, they could simply wait for the results of an imminent test series at the Marshall Islands: "It will die a natural death with the coming tests—when those tests fail."

Alvarez told Oppenheimer that there was no point in arguing, as long as they agreed to leave the program alone. As it turned out, Operation Greenhouse—which climaxed in a thermonuclear burn on Enewetak Atoll—was a success, making the issue irrelevant. All the same, Alvarez found it hard to reconcile these words with the man who had once been eager to achieve the seemingly impossible: "If it's important and doesn't violate the second law of thermodynamics, it can probably be done." But now Oppenheimer was openly voicing his doubts.

The group's draft report, which was written primarily by Oppenheimer, was brought by courier for Alvarez to review in Pasadena. Driving to Caltech, he signed it, along with the committee members Charles Lauritsen and Robert Bacher. Wondering later why it hadn't been sent to him in Berkeley, he was "forced to the conclusion" that the situation had been engineered so that he would read it in the presence of Oppenheimer's supporters rather than with Lawrence.

A few months later, Teller asked him, "Luis, how could you have ever signed that report?" When Alvarez said that it had seemed fine, Teller told him that it essentially stated that the hydrogen bomb was impeding the development of tactical nuclear weapons: "It has caused me no end of trouble at Los Alamos." Chagrined that he had misunderstood the report, which he had assumed would just be filed away, Alvarez sensed that he had been included solely to make the committee seem balanced. "Oppenheimer was, in effect, practicing Political Science 4B, on the post-graduate level, while I was still on Political Science 1A, the freshman course."

Alvarez later became exquisitely sensitive to the rhetorical power of such reports, but for now, he was painfully aware that he had been outmaneuvered. Brooding over Oppenheimer's motives, he decided to respond in kind. In May 1951, he met in Washington with Paul Fine, an AEC physicist associated with Alvarez's longtime colleague Kenneth Pitzer, whose doubts over Oppenheimer's loyalty would soon be a matter of record. Alvarez also spoke with Thomas K. Finletter, the secretary of the air force, about the possibility that Oppenheimer might be a security risk. Reviewing Oppenheimer's FBI file, Finletter quietly canceled his air force clearance, telling subordinates to avoid using him as a consultant.

A year later, Lawrence told Teller that he hoped that Arthur Compton would replace Oppenheimer on the General Advisory Committee, with Alvarez joining as well. In April 1952, Alvarez saw Finletter again at the Pentagon, along with the science advisors David Griggs and Robert LeBaron, who were close to Lewis Strauss. Recalling how Oppenheimer had used the prospect of a fusion bomb to recruit him to Los Alamos, Alvarez—who also met separately with LeBaron and Griggs—said that Oppie's prewar leftism and newfound opposition to thermonuclear weapons made it inadvisable to reappoint him. He followed up with a similar meeting with the FBI, where he seemed "agitated," but revealed nothing that they didn't already know.

Oppenheimer decided not to contest the situation, sensing that the tide had turned in favor of the three "supermen," whose ultimate victory was just around the corner. Shortly after noon in Berkeley on October 31, 1952, Lawrence entered Alvarez's office to tell him that Ivy Mike, the first thermonuclear device, had been detonated at the Marshall Islands. "Edward and I have a date to watch the seismograph needle move when the signal gets here. Why don't you come along?"

The three of them drove to campus, where they gathered in the darkened basement of the geology building to watch a point of light on a revolving drum. They estimated that it would take around twenty minutes for the primary seismic wave to arrive. Finally, the needle moved about a centimeter. Consulting a slide rule, Teller announced, "Ten megatons."

On the way back to the lab, Teller apologized for the size of the explosion. To justify the immense cost of the program, they had wanted to maximize the yield, which had been even higher than expected—four hundred times greater than the Trinity test. He was secretly exultant, and to share the news, he dispatched a triumphant telegram to Los Alamos: "It's a boy."

ON FEBRUARY 3, 1950, newspapers across the world had reported the arrest of the British physicist Klaus Fuchs, who was charged with passing vital secrets about the atomic bomb to the Soviets. Like most Americans, Alvarez learned about the scandal from the headlines, and he later said that he had encountered Fuchs only in passing: "I nodded to him in the halls when we passed in Los Alamos. I had no scientific business with him. He was a very retiring person."

Alvarez had no doubt, however, of the value of the classified information that Fuchs had shared. By enabling the Russian scientists to skip intermediate steps and dead ends, he "multiplied the effective manpower" of the Soviet program. "There are no secrets in nature, but the numbers that Klaus Fuchs and his counterparts provided the Russians certainly advanced the first Soviet nuclear explosion by several years."

William Borden was equally concerned. The former aide to Senator McMahon had grown distrustful of Oppenheimer. After learning that a communist sympathizer, Haakon "Hoke" Chevalier, had unsuccessfully tried to recruit Oppenheimer during the war, Borden wondered if there might be more to the story. He also agreed with Strauss that Oppenheimer had used the tactical nuclear weapons program to undermine the hydrogen bomb—which was exactly what Alvarez had already concluded.

In April 1953, Borden showed an unidentified "paper"—possibly an early attempt to lay out the security case against Oppenheimer—to Strauss, who moved to alert his contacts. Over lunch at Rockefeller Center, Strauss evidently gave Alvarez a preview of what was coming. Six months later, Borden sent a letter to J. Edgar Hoover, drawing on information from a security file provided by Strauss, who had been named the chairman of the AEC. "More probably than not," Borden said, "J. Robert Oppenheimer is an agent of the Soviet Union."

At the end of the year, Oppenheimer learned from Strauss that his clearance had been suspended. Instead of stepping back, he requested a hearing, which only played into the hands of his enemies. As his biographers Kai Bird and Martin J. Sherwin noted, he was clearly all too willing to question official policy: "To foreclose that possibility, [Strauss] proceeded to orchestrate a 'star chamber' hearing guided by rules that would assure the elimination of Oppenheimer's influence."

The hearing, as defined in a letter by Kenneth D. Nichols, the general manager of the AEC, was broad enough in scope to encompass Oppenheimer's opposition to the thermonuclear program: "It was further reported that you were instrumental in persuading other outstanding scientists not to work on the hydrogen bomb project, and that the opposition to the hydrogen bomb, of which you are the most experienced, most powerful, and the most effective member, has definitely slowed down its development." And if Teller was the obvious witness on this point, the next two names on the list would be Lawrence and Alvarez.

A month before the hearing, in March 1954, Alvarez met twice in Berkeley and Livermore with Roger Robb, the former prosecutor handpicked by Strauss to serve as special counsel, and his assistant,

Arthur Rolander. Alvarez was openly eager to provide ammunition. He blamed Oppenheimer for the failure of the heavy reactor project and the Materials Testing Accelerator, complaining that the lab was now "on the black list" with many physicists.

In his notes, Rolander wrote, "Alvarez said that if a star basketball player suddenly started to miss shots as Oppenheimer did in this instance, everybody would think there was something wrong." At one point, Alvarez went further, saying that the head of British intelligence had told him that "Oppenheimer was a Russian agent, worse than Fuchs." This allegation may have been too incendiary even for Robb, who wanted to focus on the thermonuclear angle.

Alvarez agreed to read his diary into the record, and Lawrence and Teller were asked to testify as well. Shortly afterward, they all met in Livermore, where Teller found Lawrence "furious to explain how dangerous Oppenheimer is." Much of what the others said was new to Teller: "Both men pointed out the thorough involvement Oppenheimer had had with communists. No one could dispute that the extent was unusual. They also pointed out the several decisions that had worked against national security in which he had played a leading role."

Teller was privately eager to sideline Oppenheimer, as was Alvarez, who was closely aligned with Strauss. According to Nuel Pharr Davis, "one of the leaders in the atomic establishment" paraphrased a statement by Alvarez as follows: "Oppenheimer and I often have the same facts on a question and come to opposing decisions—he to one, I to another. Oppenheimer has high intelligence. He can't be analyzing and interpreting the facts wrong. I have high intelligence. I can't be wrong. So with Oppenheimer it must be insincerity, bad faith—perhaps treason?"

Whether or not Alvarez truly believed this, the implication was enough. He had identified himself with the hydrogen bomb, and for his own peace of mind, he needed to neutralize the man who rejected it as a tool of genocide. Faced with a fundamental disagreement with a colleague of genius, it was easier to believe that Oppenheimer was compromised. As Lawrence and Alvarez both concluded, according to Teller, "a person of that kind cannot be cleared."

On April 12, 1954, the Oppenheimer hearing began in Washington, DC, chaired by Gordon Gray, the president of the University of North Carolina. Oppenheimer's attorney, Lloyd K. Garrison, submitted a witness list, but Robb informed him that the government wouldn't provide names in advance. As Davis later implied, however, it wasn't hard to guess who might be called: "Teller was most famous, but Alvarez . . . was probably most trusted by other physicists."

Alvarez spoke by phone with Nichols on Friday, April 23, confirming that he would testify. The following Tuesday—two days before he was scheduled to appear—Lawrence called unexpectedly from Oak Ridge, where he was attending a meeting of AEC lab directors. Outside the confines of Berkeley, Alvarez remembered, Lawrence was shocked by what scientists were saying about the hearing: "I had never heard Ernest in such an emotional state."

Loomis had already expressed concerns, and Lawrence found himself under fire from Rabi and other associates. After suffering a colitis attack, he decided not to appear as a witness, showing the blood in his toilet bowl to three colleagues to prove that it was for health reasons. On the phone, he advised Alvarez to stay home as well: "He said people had convinced him that the Radiation Laboratory would be greatly harmed if he testified and that he, Ken Pitzer, Wendell Latimer, and I were viewed as a [Berkeley] cabal bent on destroying Robert."

Alvarez understood. If the lab was damaged in removing Oppenheimer, it would be a hollow victory, and since they were likely to win anyway, it was best to avoid public association with the outcome. Canceling his plane tickets, he cabled Nichols to regretfully withdraw from "my commitment to you," followed by a phone call to confirm that neither he nor Lawrence would appear. "[Nichols] responded with disappointment but didn't ask me to change my mind. I had dreaded testifying and was pleased that I could forget the matter."

At six o'clock, however, he received a call at home from Strauss, who "let me have it with both barrels." Alvarez responded that he worked for Lawrence, whose judgment he trusted. "I had a duty to serve my country, Lewis countered. I said I had served my country during the war. Lewis's emotional intensity increased as he ran out of arguments. As a parting shot he prophesied that if I didn't come to Washington the next day I wouldn't be able to look myself in the mirror for the rest of my life. I had never disobeyed Ernest's direct orders, I said, and I wasn't about to start now."

Hanging up, Alvarez had dinner with his family. Then he sat down in the chair where he held his weekly meditation sessions. He remained there for the next two hours. It was the place where he let his imagination rove, looking for ways to use what he had recently learned. And while he never disclosed all the thoughts that passed through his mind that night, he evidently did much the same for the lingering problem of Oppenheimer—and Strauss.

The irony was that after all his efforts to diminish Oppenheimer's influence, the decisive moment had come when he was ready to move on. For the last year, he had been developing the hydrogen bubble chamber, a rewarding project that marked his spectacular return to physics. It depended on Lawrence's support, but also on AEC funding, which Strauss was in a position to control. In deciding whether to appear, he would be obeying one and defying the other.

While Alvarez was closer to Lawrence, he knew that Strauss was more dangerous—and he felt personally vulnerable to the kinds of questions that the AEC chairman could raise. After the war, Alvarez had incurred official disapproval over a diary that he kept on Tinian, as well as pictures "relating to the container of some part of the bomb." The diary had been examined and returned, but the photos were a different story. According to a security officer, Alvarez had been "somewhat hard-headed and reluctant to give up the pictures until admonished."

He was criticized afterward as "a little dubious and resentful concerning security at Los Alamos," and his judgment would be questioned again. On March 20, 1948, the *Chicago Tribune* quoted him as stating that radar crews in Korea had seen Russian aircraft flying at supersonic speeds. Alarmed, the air force asked the FBI to investigate whether he had "intended to harm the United States or to benefit a foreign country." Alvarez explained that he had heard the rumors from operators of his Microwave Early Warning system. He repeated them casually to Lawrence's friend Robert McCormick, the owner of the paper, who printed them without permission.

Alvarez was "annoyed and embarrassed" by the incident, and he probably never knew about the investigation of an even less dignified episode in 1951. On a family ski trip to Sun Valley, Idaho, Alvarez had broken his leg. Striking up a conversation in the hotel lobby with a vacationing couple, he noticed the husband's naval ring and launched into a loud discussion—without any prompting—of his classified work. He seemed eager to be seen as "quite the big shot," and the recipient of his attentions was disturbed enough to report it. An inquiry into his "bragging campaign" concluded that it reflected "a certain lack of discretion."

But he would have been most concerned by his association with the Chevalier affair, the cornerstone of the case against Oppenheimer. In 1942, Peter Ivanov, a Soviet intelligence officer at the San Francisco consulate, had approached George Eltenton, a Communist Party member who worked as a chemist at Shell Oil. Ivanov asked if he knew anything about war research at Berkeley, specifically mentioning Oppenheimer, Lawrence, and a man Eltenton described as "a third party whose name I do not recall," although he later said that it might have been Alvarez.

To sound out Oppenheimer, Eltenton asked their mutual friend, Hoke Chevalier, who taught French literature at the university. Later that winter, Chevalier dined at Oppenheimer's house, where he raised the possibility of passing information to Eltenton's contact. Oppenheimer refused, calling it "treason." About six months later, he shared his concerns about Eltenton with Colonel Boris Pash, the West Coast head of army intelligence, implying that attempts had been made to recruit others: "I have known of two or three cases, and I think two of the men were with me at Los Alamos."

Oppenheimer subsequently dismissed his reference to multiple approaches as a "cock and bull story" that he invented to protect

Chevalier—a rationale that even his defenders found incomprehensible. As his biographers Bird and Sherwin noted, however, Ivanov had reportedly asked about Oppenheimer, Lawrence, and possibly Alvarez. "It seems entirely reasonable to suppose," Bird and Sherwin wrote, "that Eltenton would have mentioned the three names to Chevalier—and that Chevalier may well have specifically mentioned them to Oppenheimer."

At the time, Alvarez was at MIT, but he was a perfectly plausible figure for an outsider to bring up in connection with war work at Berkeley. In any case, Eltenton's statement—which ranked Alvarez alongside Oppenheimer and Lawrence as a potential target—was enough to put the intelligence community on alert. And it would feed indirectly into the strangest incident of all, in which Alvarez briefly became a leading suspect in the investigation of espionage at Los Alamos.

In 1943, the classified Venona project was established to decrypt Soviet diplomatic cables intercepted during World War II, producing a trove of material that led to the identification of Fuchs, among others, as a Russian spy. References to covert operatives were usually by code name, or cryptonym, and immense resources were devoted to uncovering their identities.

One cryptonym was "Shmel," which referred to an employee at Los Alamos who was recruited as an informant at the end of 1944. Five years later, the FBI field office in Albuquerque conducted a search of personnel records for members of the Manhattan Project by the name of Shmel, "possibly phonetic." In less than a month, the investigators were focusing on Alvarez, whose middle name was given on his birth certificate—according to the Albuquerque office—as "Schmell [*sic*]."

The report speculated, incorrectly, that it was the maiden name of Alvarez's mother, and no one ever connected it to his father's former mentor, Dr. Emile Schmoll. In March 1950, Alvarez was described as "the best suspect developed to date." To underline his significance, an investigator cited rumors of a wartime plot by the Soviets to "kidnap certain key scientific personnel." Two of the targets were Oppenheimer and Lawrence, while the third was a scientist—tentatively identified as Alvarez—who had invented a "new circle."

Eventually, the investigation concluded that Alvarez's travels didn't fit with "Shmel," who turned out to be David Greenglass, the brother of Ethel Rosenberg. Greenglass, who worked as a machinist on molds for explosive lenses, once pocketed one of Alvarez's exploding bridgewires, passing it along to a Soviet contact. The FBI was unaware of this, but it knew that he had given a sketch of the implosion mechanism—possibly the "new circle"—to his brother-in-law Julius. Greenglass, whose testimony sent the Rosenbergs to the electric chair, had been called "Shmel," or "bumblebee." It was all just an incredible coincidence.

Alvarez probably never met Greenglass, but he had reason to be worried by his apparent proximity to other confirmed spies. In the period when he was still seen as the best suspect for "Shmel," an AEC security officer darkly noted "that among the close associates of Fuchs at Los Alamos was Luis Alvarez." Even more troubling was Ted Hall, the young physicist whom Alvarez destroyed at billiards. While Hall was building ionizing chambers for the RaLa experiment, he was also delivering documents to the Soviets, as the Venona project had determined by 1950.

As AEC chairman, Strauss had access to much of this information. Although he was never questioned about it on the record, Alvarez knew that Strauss could be a formidable enemy, whether by holding back funds or by encouraging doubts about his loyalty. As he considered the situation, it became all too clear that someone from Berkeley needed to testify. A vengeful Strauss could cause more damage than all the recriminations of the scientific community, and if Lawrence refused to go, Alvarez was willing to take it upon himself.

In his memoirs, Alvarez described it rather differently: "I finally concluded that I really would be ashamed to think that I'd been intimidated. So I poured myself a stiff drink, booked a seat on the TWA midnight red-eye flight, and sent Lewis a telegram. Later that evening I drove to the airport to disobey Ernest for the first time and to give testimony that might hurt a friend, though I hoped it wouldn't." In a line cut from the published version, he continued, "I also knew that my action would be unpopular with my scientific friends. This was a dangerous moment in my life."

He arrived in Washington early on April 28. That evening, he dined with Roger Robb at the Chevy Chase Club in Maryland, where the special counsel prepared him for his appearance, as he had already done for Teller. A member of Oppenheimer's legal team, Allan Ecker, recalled that Robb played a recording of his client's interrogation by Pash one night for "people who were afterwards to be witnesses." Whether or not Alvarez heard it, he may well have learned of Oppenheimer's reference to "two or three cases," and perhaps of the possibility that his name had come up.

Alvarez was called to testify on the afternoon of Thursday, April 29, 1954. The location was a temporary AEC building at Sixteenth Street and Constitution Avenue, close to the center of the National Mall. He entered Room 2022 to find a long, dim space with the members of the board—Gordon Gray, Ward V. Evans, and Thomas A. Morgan—seated at one end. Perpendicular to the head table, the attorneys for the opposite sides sat across from one another, with a chair for witnesses positioned to face the board members. Behind it was a leather couch on which Oppenheimer smoked in silence throughout Alvarez's testimony.

After Alvarez was sworn in, he calmly answered questions from Robb, with the occasional interjection from Lloyd K. Garrison. Describing his background and his role in the Manhattan Project, he emphasized that Oppenheimer had used the possibility of a fusion bomb to convince him to join: "He certainly raised no question about the morality of the thing."

His diary from October 1949, which Oppenheimer's team was unable to review in advance, was read into the record, with Alvarez expanding on it as needed. He said that he had assumed that Oppenheimer would be "enthusiastic" about the thermonuclear program, recounting his shock on learning otherwise: "The program that we were planning to start was not one that the top man in the scientific department of the AEC wanted to have done."

Robb asked if the hydrogen bomb could have been developed sooner if it had been given a higher priority. "I think brilliant inventions come from a concentrated effort on a program," Alvarez said. "The reason there were not any brilliant inventions in the thermonuclear program for four years after the war is that there was no climate to develop in. Lots of people were not thinking about the program. Essentially one man was, and it is very hard to generate ideas in a vacuum."

They ended shortly before six, resuming at nine thirty the next morning. Robb opened with one last line of questioning. Referring to Alvarez's discussions with scientists about the hydrogen bomb in October 1949, Robb asked, "At that time these individuals were enthusiastic for going ahead with it—is that right?"

Alvarez understood where this was headed. "That was my very strong impression."

"To your knowledge," Robb continued, "were those conversations in advance of any talks that these people had with Dr. Oppenheimer?"

"I think that is so, sir," Alvarez responded. "I am sure it is so in the case of Dr. Serber. I am quite sure in the case of Drs. DuBridge and Bacher, and also in the case of Dr. Rabi."

Robb came to the crucial question. "Subsequently these people changed their views—is that right?"

"Quite drastically, yes," Alvarez said. As Robb wound down the direct examination, they knew that they had left the impression that Oppenheimer had convinced these scientists to change their minds.

The questioning was taken up by one of Oppenheimer's attorneys, Samuel J. Silverman, who did what he could to weaken Alvarez's points. He noted that the General Advisory Committee had been in favor of a reactor that could produce excess neutrons, which cut against the argument that it had deliberately derailed the heavy water pile at Berkeley. "Being in favor of piles is like being against sin," Alvarez replied. "I think everyone is for piles, but nonetheless none got built."

When asked about the Long Range Planning Committee report, which had supposedly been used to undermine the thermonuclear program, Alvarez said that the hydrogen bomb had been "more or less damned by faint praise." Stating that the impact would be obvious only to someone who understood the consequences of policy papers, he added, "I would certainly say that Dr. Oppenheimer is one of the most skilled document writers that I have ever run across."

As Silverman tried to challenge the implication that Oppenheimer had engineered the report to undercut the hydrogen bomb, Robb objected that the questions were "frivolous." Silverman pushed back: "There is nothing frivolous about them. Here is a man that signed the report and didn't know what was in it, although he was the representative of the opposite camp on that precise point."

To frame the document as a strategic coup by Oppenheimer, Alvarez had to admit to his own carelessness. "I have reread the report, and knowing now what happened at Los Alamos, I can see why it happened, and I can see that I was not careful enough to guard against this possibility." Acknowledging his "lack of vigilance," he agreed when Silverman asked if he "fell down on the job." He concluded, "I signed the document, which I thought fairly reflected the views which I heard expressed in the meeting. I found out later that I had been had."

On Vannevar Bush's alleged statement that Truman hadn't trusted Oppenheimer, Alvarez said, "Until that time I had always thought that Dr. Oppenheimer was the most loyal person, the most wonderful man. He is one of my scientific heroes." At the same time, he characterized Oppenheimer's role in the opposition to the hydrogen bomb in vaguely sinister terms: "Every time I have found a person who felt this way, I have seen Dr. Oppenheimer's influence on that person's mind."

Describing Oppenheimer as "one of the most persuasive men that has ever lived," Alvarez granted that his actions didn't mean that he was disloyal, but maintained that he was wrong about the thermonuclear program: "He showed exceedingly poor judgment. I told him so the first time he told me he was opposed to it. I have continued to think so." Evoking the prospect—recently fulfilled—of a successful Russian thermonuclear test, he said, "I felt sure that this would be one of the most disastrous things that could possibly happen to this country."

Alvarez was careful to end on a note of praise. "I don't want in any way to minimize Dr. Oppenheimer's contribution, because to my way of thinking he did a truly outstanding job at Los Alamos. I think he was one of the greatest directors of a military program that this country has ever seen." After a few more questions, he was done. Later that day, Strauss saw him briefly, confirming that they were on good terms again, and Alvarez felt that Lawrence secretly approved as well.

The hearing concluded on May 6, followed three weeks later by a formal verdict. In a vote of two to one—Evans, the only scientist, was the lone dissenter—it found that Oppenheimer was a "loyal citizen" but recommended against renewing his clearance. The majority stated, "We find his conduct in the hydrogen-bomb program sufficiently disturbing as to raise a doubt as to whether his future participation . . . would be clearly consistent with the best interests of security."

Although the witnesses had been assured that their testimony would be confidential, the entire transcript was published in June, shortly before Oppenheimer's clearance was officially revoked. The first to feel the blowback was Teller, who had said that he would prefer "the vital interests of this country in hands which I understand better." Alvarez lamented Teller's "rough treatment" to Strauss, who recalled, "He said Dr. Rabi was there the day before yesterday and refused to shake hands with [Teller]. Alvarez said he thought Dr. Teller was very low in his mind and it would be an act of kindness for me to call him and encourage him."

While he and Teller would be widely described as "notable exceptions" to the scientists who rose to Oppenheimer's defense, Alvarez felt that he had avoided similar shows of hostility. By stating his admiration for Oppenheimer, he believed that he had somehow threaded the needle, although he later granted, "All of us who were touched by the hearing were in some way wounded. No one has ever directly criticized me for testifying as a government, though not unfriendly, witness. I've felt disapproval but it hasn't been expressed, nor am I aware of having lost any friends."

Alvarez suspected that people whispered about his testimony behind his back, but he claimed, notably, that he never brought it up with

Oppenheimer himself. In his memoirs, he wrote, "In the years after the hearing, Robert and I had a number of pleasant conversations at various conferences, about physics and the old days at Berkeley. Our long friendship was equal to the task of ignoring the hurts we had both experienced along the way to doing what we saw as our duty. But neither of us ever referred to the painful hearing days."

Elsewhere, however, in an interview that he gave over a decade after Oppenheimer's public humiliation, Alvarez indicated that the damage was real and lingering. "Oppie and I meet once a year at the High Energy Physics Conference, held in Rochester. The first year after the hearing, Oppie didn't speak to me—very pointedly ignored me. The second year he spoke to me, but his wife snubbed me, and the third year Oppie and his wife were both cordial. Since then they acted as if nothing had happened—although we know that something did happen."

6.

BUBBLE CHAMBER

1950–1963

IN THE YEARS THAT HE LOST TO THE MATERIALS TESTING Accelerator, Alvarez suffered what he later called a "midlife crisis." On returning to Berkeley from Livermore, he avoided having lunch with the other physicists, whose work he was unable to understand. Instead, he played endless games of hearts with the younger technicians who analyzed nuclear emulsions. "It never occurred to me that I was really hiding out from my physics friends, afraid of exposing my ignorance."

The aura of failure from the MTA sidetracked him at a moment that he described afterward as the dawn of "modern particle physics." In 1946, three Italian physicists had demonstrated that the mesotron, which was widely assumed to be the carrier particle for the strong force that held together the nucleus, was really nothing of the kind. It was actually a negative "heavy electron," later known as the muon, a result so unexpected that Rabi quipped, "Who ordered that?"

Physicists had been denied their carrier particle, but a year later, the pion—observed in emulsion plates exposed to cosmic rays—fulfilled all the necessary requirements. As scientists moved beyond the nucleus to the mysterious products formed by collisions at high energies, the unexplored frontier demanded a corresponding shift in technology. Pions were highly unstable, decaying into muons almost immediately, so the next step was to produce them in the lab.

Lawrence authorized a search at the 184-inch cyclotron, which yielded pion tracks in 1948. More surprises followed, as cosmic rays disclosed "strange particles," with vanishingly brief lifetimes, that still survived for a trillion times longer than predicted. They were apparently produced by the strong force, but no one understood why it didn't break them down again at once.

At Caltech, Murray Gell-Mann conceived of a property that was conserved by the strong force, preserving the particles that it created, but not by the weak force responsible for radioactive decay, which destroyed them a hundredth of a microsecond later. He called it "strangeness." It was a revolutionary proposal that underlined the relationship between theory and experiment, as well as the need for novel thinking in a field that was desperate for new ideas.

Throughout it all, Alvarez wasn't involved, and he was pained by the knowledge that his peers had made strides without him. Lawrence, who had received the Nobel Prize in 1939, was followed over the next decade and a half by Rabi, McMillan, Seaborg, and Felix Bloch. As an anonymous associate told the author Nuel Pharr Davis, "[Alvarez] was the only one of those sunbathers who never got a Nobel Prize, and they kept him reminded."

Alvarez heard that McMillan and Seaborg had won the prize over a radio broadcast while driving to work. At the lab, he congratulated McMillan and asked the switchboard operator to announce it on the intercom. "And then I suddenly felt an emotion I've never felt before, that hit me in the stomach, and it could most probably be described as jealousy, although I am sure that I was unconsciously worried that Ed's prize might affect our relationship, which had been so rewarding to me in the past. I went home and took an hour's nap, and came back completely recovered, and I have never had another symptom of this kind."

In reality, such signs of success meant everything to Alvarez, who was no longer even the most famous man in his family. His father, Walter, had been forced into retirement by the Mayo Clinic, where he was seen as an "enfant terrible" who disregarded professional boundaries: "I paid a price for my independence." On returning to Chicago, he began writing a popular health column that was syndicated in newspapers worldwide. Before long, his son was frequently asked whether he was related to "Dr. Alvarez," who became known as "America's family doctor."

Alvarez was tempted to make a similar career change. As he hesitated over a return to accelerator physics, "where I had demonstrated some talent," he was still in demand as a corporate and government advisor. After the war, he had consulted for Project RAND, a defense think tank founded at Douglas Aircraft. Over a decade before Sputnik, Alvarez participated in a landmark study on artificial satellites—he offered advice on radar and nuclear propulsion—and he continued to work with the RAND Corporation for the rest of his life.

In the summer of 1950, he joined Project Hartwell, a navy committee that evaluated the threat to shipping from Soviet submarines. Alvarez and the other members recommended implementing a passive sonar tracking system, with subs detected by an array of hydrophones on the seabed. Jerrold R. Zacharias, the MIT professor who directed the study, recalled how Alvarez refused to retreat in the face of opposition: "In his own way, I think, he tries to be reasonable. But he has very strong opinions, and I think it is his arrogance that bothers me the most."

Alvarez enjoyed being treated as an expert. His most valuable contact was the lawyer and venture investor Rowan Gaither, who oversaw RAND's evolution into an independent nonprofit. During the war, Gaither served as a liaison between manufacturers and the MIT Rad Lab, although he left Alvarez "singularly unimpressed" at their first meeting. Over the following decade, however, Gaither closed a deal to build equipment for blind landings; procured radar sets for the linear accelerator; and, as president of the Ford Foundation, arranged for Alvarez to buy a Lincoln convertible every year at factory cost. Eventually, he came to see Gaither as his "closest friend."

In 1951, Gaither sponsored Alvarez's membership in the Bohemian Club. The notoriously private organization—entirely male and overwhelmingly white—was best known for its exclusive campground in a redwood grove in Monte Rio, California. In the early days of the Manhattan Project, Lawrence and Oppenheimer had hosted a planning session there on magnetic isotope separation, and it remained an elite location for networking at the highest levels.

The club became enormously important to Alvarez, who had first been invited there as Lawrence's guest. At its annual gathering at Bohemian Grove, two thousand members assembled for talks, amateur theatricals, and informal dealmaking. It was there that he befriended Bill Hewlett and Dave Packard, who installed him on their company's board, and remained close to Lewis Strauss. He relished mingling with the powerful, although he felt that he initially owed it to his more influential campmates: "I had been welcome because of the friends I accompanied."

Lawrence himself was going through a difficult period. In 1949, after he and Alvarez saw a crude demonstration of color television, Lawrence decided that he could do better. Alvarez—who secretly tested the display screen of the prototype set with a magnet in his pocket—was doubtful of the results: "Ed McMillan and I weren't impressed with Ernest's primitive and unworkable color tube. [Lawrence] felt otherwise and without much difficulty secured extensive financial backing from Paramount Pictures."

In partnership with Gaither, Lawrence hired Alvarez and McMillan as consultants for Chromatic Television Laboratories. Working out of the garage of his vacation home, he built prototypes using staff and materials from Berkeley, which struck Alvarez as ethically questionable. Although Alvarez assisted with demonstrations and lent his name to a patent, he was frustrated by how television dominated his conversations with Lawrence: "Ed McMillan and I were quite disgusted about this, but we couldn't do anything about it."

For Alvarez, it became a cautionary tale. Lawrence framed television as a hobby—"Some people like to play bridge, some people like to work crossword puzzles"—but it left him increasingly isolated. "He got to the point where he couldn't talk physics with anybody outside of the laboratory, because he was, in a sense, a little afraid to show his ignorance. I was one of the few people who could talk modern physics with him, because I knew what he didn't know."

Alvarez had narrowly avoided the same fate. In the fall of 1952, after his return to Berkeley, he had a casual conversation with Herbert York, whom Lawrence had chosen to run the new weapons lab in Livermore.

York asked Alvarez to assume responsibility for his two best graduate students. Sensing an opportunity in an otherwise routine request, Alvarez agreed on one condition: "I would hire them as my research assistants if they would treat me as *their* research assistant."

To get back into the game, Alvarez wanted the students, Frank Crawford and Lynn Stevenson, to assign him homework on particle physics. According to Stevenson, it took just "a few microseconds" for Alvarez to catch up: "What actually happened . . . was a veritable torrent of ideas from Luie, which kept us going literally day and night." His first experiment—a search for pions in the nucleus—came up empty, but he felt that he had finally been "rehabilitated."

As an inventor, Alvarez had conceived of a promising idea the year before, after reviewing his lecture notes over breakfast. Driving to campus, he suddenly pictured a new kind of linear accelerator. An electric terminal at the halfway point would attract negative hydrogen ions, which were stripped of electrons by a foil sheet, turning them into positive protons that automatically received a second push. Since it would double the beam's energy at no cost, Alvarez called it the "swindletron." Rebranded as the tandem accelerator, it was engineered by John Trump, whose nephew Donald would later speak proudly of his uncle's genius for physics.

In April 1953, Alvarez attended the annual American Physical Society meeting in Washington, DC, where he hoped to publicize another invention. To improve his golf game, he put together a training system with five stroboscopic lamps. As he swung the club, it set off a series of photocells, producing a succession of sharp images that enabled him to refine his form. Lewis Strauss arranged for him to present one to President Eisenhower, who spent fifteen minutes playing with the golf trainer before it was installed in the White House gym.

The conference was more significant for other reasons. At the Shoreham Hotel, Alvarez was at a table in the garden, reminiscing with friends from the war, when he noticed a young man seated nearby. It was Donald Glaser of the University of Michigan, who had a presentation scheduled for the last day—"the crackpot session"—when few people would be in the audience. Confessing that he wasn't planning to attend, Alvarez asked about his work.

Glaser said that he was developing a new particle detector—the bubble chamber. Although a myth later arose that he had been inspired by the bubbles in a glass of beer, he had actually made a methodical list of all the instabilities that could be catalyzed by a charged particle. Knowing that a stray cosmic ray could cause a very thin stream of liquid to vaporize, he decided to put a similar principle to use.

The result was a small glass chamber—about an inch long—filled with ether, which he heated to just short of boiling. By decreasing the pressure with a piston, he created an unstable state that could be vaporized by the smallest disruption. When a positive or negative particle entered, it produced an ionization track lined by pairs of oppositely charged molecules. The heat caused the ether to boil along its path, leaving a visual record in a trail of tiny bubbles.

Examining the track photos, Alvarez grew excited. He was already thinking about the next stage in physics, which lay in the massive proton synchrotron under construction in Berkeley. Lawrence had never given up hope of detecting the antiproton, to the point that Alvarez and McMillan had quietly formed "a protective unit" to prevent him from announcing it prematurely. To realize his dream, Lawrence authorized an accelerator that would yield energies of billions of electron volts, inspiring its futuristic name—the Bevatron.

Alvarez thought that it would be wasted on the current generation of detectors. With their low density, cloud chambers missed too many events. Nuclear emulsion plates had excellent resolution in space but not time, limiting their usefulness for reactions that involved neutral particles—which left no visible trace—in their intermediate steps. What was needed, he concluded, was a detector with high density that could be quickly reset in time for the next pulse.

As a result, he was even more prepared than Glaser to see the bubble chamber's potential. At his hotel that night, he discussed it with his colleagues. Instead of a hydrocarbon molecule, like ether, which he compared to "a ball of junk," his version would use liquid hydrogen, allowing collisions to occur directly on protons. Like the cyclotron, it would start small and gradually ramp up in size.

Back in Berkeley, Crawford and Stevenson began by attempting to duplicate Glaser's work with ether, but the glass chambers were frustratingly hard to keep clean. In the fall, John Wood, a technician, went ahead and built a chamber—an inch and a half in diameter—with liquid hydrogen. Stray bubbles appeared at the edges, but it still produced good images.

Looking at the results, Alvarez realized that the engineering would be simpler than he had feared. Because superheated liquids boiled at tiny imperfections on the inner walls, known as nucleation sites, he had assumed that the inside surface needed to be perfectly smooth. Now he saw that it didn't matter if bubbles formed there, as long as the chamber could be rapidly set up and refreshed—and "dirty" metal chambers with glass windows would be much easier to build.

This insight also fundamentally altered the nature of the tool itself. Glaser had hoped that scientists could use it to research cosmic rays, but a dirty chamber had to be cycled every few seconds. It would be perfect for the Bevatron, where it could be timed to coincide with the extraction of the beam, but not for sources that couldn't be controlled. "I was trapped," Glaser recalled. "It was just what the accelerators needed and it wasn't useful for cosmic rays."

When the Bevatron became operational, Alvarez devoted himself to mastering its intricacies. One night, a technician pulled too hard on the target probe, accidentally yanking it out completely. As air rushed into the vacuum tank, Alvarez pressed his palm over the opening, giving the crew a chance to close the chamber. "The Bevatron was able to pump back down without excessive loss of time," an onlooker remembered, "while Luie was rubbing the sore spot on his hand."

The Bevatron turned out to be an ideal factory for producing strange particles. With Crawford and Stevenson, Alvarez used it to study the K meson, or kaon, which seemed to come in more than one variety. Measuring thousands of oscilloscope photographs, they confirmed that the kaons were identical in most respects, but somehow decayed in different ways. "Don't you worry about it," Lawrence correctly predicted. "The theorists will find a way to make them all the same."

On April 29, 1954, while Alvarez was at the Oppenheimer hearing, his team obtained pictures from a 2.5-inch bubble chamber. They were already looking ahead to four inches, then ten, but he wanted to go even further. To study strange particles, they needed to maximize the odds that a pion beam would collide with a proton; make the chamber long enough for products to curve in a magnetic field, which was used to calculate the momentum and sign of charge; and enable reactions involving neutral particles to play out until the tracks reappeared.

"The result of this straightforward analysis was a rather frightening set of numbers," Alvarez said. "The chamber was to be 72 inches long, 20 inches wide, and 15 inches deep." Its magnet alone would need two or three megawatts of power and weigh at least a hundred tons. The window—the largest piece of optical glass ever cast—would be exposed to enormous pressures, with a huge volume of liquid hydrogen that could explode catastrophically in an accident.

In April 1955, Alvarez—now an assistant director at the lab—prepared a proposal for the seventy-two-inch chamber. Advances in both accelerators and detectors were crucial to physics, he wrote, "but since the accelerators are the more expensive and photogenic objects, they are often regarded as being the more important members of the partnership." As a result, physicists at the Bevatron were stuck with the slower and less sensitive detectors meant for cosmic rays.

Alvarez pointed out that a bubble chamber with rapid cycling could yield more images in a few hours than could be processed by all the world's human technicians in a year. Based on his experience with radar tracking, as well as with the MANIAC computer at Los Alamos, he was confident that he could develop automated scanners and software to handle this flood of information. What he didn't mention was the cost, which he privately estimated at $2.5 million.

To pay for what he later compared to an "entire weapons system," he needed a special appropriation from the AEC. It was a test of his credibility, but Lawrence seemed skeptical—the largest chamber that they had built so far was only four inches in diameter. Surprised to hear doubts from a man who had spent his life scaling up existing tools, Alvarez countered that a seventy-two-inch chamber was really just a series of ten-inch chambers in a row. Unlike many design problems, where "one essential criterion can be met only if the object is very large while an equally important criterion demands that it be very small," the basic principles were independent of size.

The argument wasn't particularly convincing—even the ten-inch chamber was still on the drawing board—but Alvarez promised to return the money if he couldn't build the smaller version first. Lawrence finally agreed to help, as long as Alvarez pledged to stick with the chamber until it was finished. The remark struck him as a needless rebuke. "I do flit from one project to the next, although I like to think that I never left any of them, except the MTA, before its success was assured. Ernest clearly didn't recall my career in the same way."

In Washington, Lawrence and Alvarez spent a morning in meetings with John von Neumann, Willard Libby, and Lewis Strauss. At a cocktail party hosted that night by von Neumann, Alvarez learned that he had been given the funds. He sensed that Libby, an old friend, had persuaded the others, and he knew that Strauss might have been less helpful if he hadn't testified at the Oppenheimer hearing. As Alvarez eagerly prepared to return to the front lines, he often thought of something that Lawrence had said: "I don't believe in your big chamber, but I do believe in you."

WHILE ALVAREZ ORCHESTRATED his victorious return to physics, he continued to be in demand as a government consultant. In 1951, on his ski trip to Sun Valley, he had made his stature clear to his listeners at the hotel. The FBI report on the incident noted, "One of the unknown guests asked Alvarez about flying saucers, and Alvarez immediately talked it down, justifying his position by stating he had recently been in conference with the chief scientists of the United States Air Force who would certainly know if there was any truth to such matters."

He had reason to believe that he would be entrusted with information about unidentified flying objects. During the war, he had taken an interest in "foo fighters," the mysterious balls of light that pilots observed alongside planes. Alvarez suspected that they were an electrostatic phenomenon, like St. Elmo's fire, or reflections from airborne ice crystals. While they were never conclusively explained, he felt that they were "not beyond the domain of present knowledge of physical science."

A few years later, he heard about other sightings at an experimental station in Arcata, California, that was developing blind landing systems. In 1947, a technician named Kenneth Ehlers began recording and photographing "gizmos" on his radar scope that moved at thirty miles per hour. Alvarez speculated that they were ionized particles in the atmosphere, but he didn't know how they could become dense enough to resemble an aircraft. "And I don't understand why the gizmos don't always move with the wind, *if they aren't alive*."

Five years later, the air force responded to a national surge in flying saucer reports with the investigation known as Project Blue Book. Its director, Captain Edward J. Ruppelt, planned to present its results to a scientific panel organized by the CIA. Howard P. Robertson, a Caltech mathematician with a background in radar, chaired a group of five experts, including Alvarez, who was described as "an outstanding scientist in the fields of radar operation, characteristics, and anomalies."

Alvarez later claimed that he assumed that he was working for the air force, and he didn't remember that it was the CIA until he received his compensation check from a private bank account in Pittsburgh. In fact, he was undoubtedly aware of the panel's origins and objectives. Another member, the astronomer Thornton Page, recalled that Robertson said that "our job was to *reduce* public concern, and show that UFO reports could be explained by conventional reasoning."

The Robertson Panel held its first meeting at the National Academy of Sciences in Washington on January 14, 1953, with Alvarez concentrating on "case histories involving radar or radar and visual sightings." Its members were advised that their recommendations would determine the future of Project Blue Book. If they decided that the evidence pointed to "interplanetary spacecraft," their report would go directly to President Eisenhower.

After the lights were dimmed, two films of unexplained UFOs were projected on the conference room wall. One was taken in Great Falls, Montana, on August 15, 1950; the other, which inspired the most discussion, near Tremonton, Utah, on July 2, 1952. The latter was captured by Delbert C. Newhouse, a navy photographer, with a telephoto lens. For forty-five seconds, Newhouse filmed a dozen bright ovals in formation against the sky, keeping the camera steady before following one object that appeared to depart from the others.

According to the experts, the objects weren't "birds, balloons, or meteors," and they seemed luminous in themselves. Alvarez asked to view the footage several times, Ruppelt wrote, before advancing an explanation: "[He] thought the UFOs could be sea gulls soaring on a thermal current. He lived in Berkeley and said that he'd seen gulls high in the air over San Francisco Bay." To explain the unusual speed of the object that broke from the rest, Alvarez said that Newhouse might have "panned with the action" without realizing it. The following morning, the members watched a film of seagulls, paying close attention to how their bodies reflected the sunlight.

Over the next two days, the panel reviewed additional incidents. Alvarez discussed using radarscope cameras to capture "peculiar radar echoes," which might be caused by interference between neighboring stations. As a solution, he proposed a switch that would allow radar operators to slightly alter their own frequency. If a "gizmo" moved onscreen in response, it was an artifact, not an aircraft.

They also considered sightings accompanied by evidence in the form of radiation. In October 1949, a spike from cosmic ray detectors at Palomar Mountain, California, coincided with a V-shaped formation of saucers. The following year, a team of scientists recorded radiation data during similar events over two months. After examining reports and circuit diagrams, however, Alvarez attributed them to "instrumental effects" that a competent user would have recognized.

Alvarez was dismissive of the Robertson Panel in his memoirs: "We listened for a week while people who claimed to have seen UFOs retailed some of the wildest stories I'd ever heard. We found no threat, no phenomena attributable to foreign artifacts, and no need to revise current scientific concepts." He said much the same to Arthur C. Clarke, telling him that they could rule out most of the sightings "with a little common sense and a modicum of science."

In fact, Alvarez was misrepresenting both the panel—which heard only from expert witnesses, such as the astronomer J. Allen Hynek, and focused on incidents that Project Blue Book couldn't explain—and his own experience. He later wrote to Page about their "exciting days" together, and not every sighting could be disregarded. Alvarez and Robertson categorized at least two "foo fighter" events as "unexplained but not dangerous," which they granted was unsatisfying: "They were not happy thus to dismiss the sightings by calling them names."

By Saturday, the panel was ready to deliver its conclusions, which were narrowly tailored. The members found no evidence of "a direct physical threat to national security" or phenomena beyond the bounds of science. They were more concerned that false reports would distract from real dangers, as well as by "a morbid national psychology in which skillful hostile propaganda could induce hysterical behavior and harmful distrust of duly constituted authority."

Among its other recommendations, the panel proposed investing in security programs to remove "the aura of mystery" around flying saucers. In the aftermath, Project Blue Book shifted to downplaying sightings, which were publicly discussed only if they could be explained. Hynek—who later developed the "close encounters" system for classifying UFO contacts—was critical of the panel's attitude: "I had the distinct impression they didn't care."

Yet there was no doubt that they had addressed the problem—defusing concern over UFOs—that seemed most urgent at the time. In the past, Alvarez had used his knack for coming up with alternative explanations to deflect reasonable questions about radar and the bomb. It contrasted with his combative approach to physics, but it was perfectly consistent with a political attitude that could be accurately, if incompletely, described as "conservative."

On a deeper level, Alvarez had minimal interest in disrupting the orderly systems of power that allowed him to take risks in his own career. David E. Lilienthal—who dismissed Alvarez, Lawrence, and Teller as "scientists in gray flannel suits" with the promotional instincts of advertising men—noticed the same caution in Oppenheimer after the hearing: "I sometimes wonder whether those of us—such as myself—who have been under terrific attack don't go out of our way to be conservative in discussing the position of our country and our government."

Of scientists who served on committees, Oppenheimer said, "They like to be called in and asked for their counsel. Everybody likes to be treated as though he knew something." For Alvarez, it offered many rewards. Rather than committing himself to years of work, he could deliver an opinion and move on; he was exposed to valuable new ideas; and it was ideal training for moving between disciplines. All it required was a willingness to respect certain boundaries, which had been established beyond any doubt by Oppenheimer's excommunication.

In April 1956, Alvarez heard that a dozen American physicists would attend the Moscow Conference on High-Energy Physics, where the Soviets planned to display the most powerful cyclotron ever made. His name wasn't on the list. "I knew that someone had goofed. The Russians needed to know about liquid hydrogen bubble chambers, and I was obviously the person to give them those details." His work was a notable part of the Cold War race for technological supremacy, and he badly wanted to be among the first scientific visitors to the Soviet Union.

A week later, he was added to the roster of delegates approved by the AEC. On May 14, he arrived in Moscow with a group that included Panofsky, Gell-Mann, and his frequent antagonists Jack Steinberger and Emilio Segrè—although the main topic of conversation, the Italian physicist Bruno Pontecorvo, had yet to show his face. A devoted communist, Pontecorvo shared a patent with Fermi and Segrè on the production of slow neutrons, but he fell under suspicion after his colleague

Klaus Fuchs was arrested in 1950. Shortly afterward, he defected to the Soviet Union.

Alvarez saw that everyone was eagerly anticipating Segrè and Pontecorvo's reunion: "People were anxious to see how the two would act toward each other here." His own relationship with Segrè, already strained by the loyalty oath, had been further tested over compensation for the slow neutron patent. When Alvarez argued in Berkeley that the Italian scientists should be content with their welcome in America, Segrè had replied that he hadn't known that citizenships were for sale. Alvarez was furious, Segrè recalled, at the prospect of a lawsuit: "He concluded that I should let him know when Pontecorvo writes me from Russia."

Segrè prevailed in court, but he was reluctant to be seen with Pontecorvo, who eventually cornered him at the conference. Alvarez pointedly showed the others that he had no reason to worry. After a talk on the bubble chamber, he ran into Pontecorvo at an intermission with soda and cake. Fortunately, they had a safe subject to discuss. In 1949, Alvarez had proposed a way to detect the neutrino, the plentiful particle that passed through most matter without a trace. Using a giant tank of carbon tetrachloride—or dry-cleaning fluid—as a target, they could measure the radioactive argon produced by the extremely rare absorption of neutrinos by chlorine-37.

In Moscow, Alvarez told Pontecorvo, who had explored a similar idea, that the chemist Ray Davis was pursuing it for real at Brookhaven National Laboratory in New York. It gave them just enough material for a brief but cordial interaction. "Pontecorvo was very interested in the results, but was obviously in a hurry to get away, before we ran out of physics talk."

The main attraction was the Moscow cyclotron. Stunned by its workmanship, Alvarez made a detailed sketch from memory as soon as he could. At their next stop, he saw the world's largest synchrotron nearing completion, although he noticed a lack of space for detectors: "I wouldn't be surprised to find the wall on one side torn down the next time we come."

He observed similar blind spots throughout his visit. The Soviets built beautiful machines, but they didn't know how to use them—they were just repeating experiments that had been done elsewhere. In a severely hierarchical system, he concluded, it was hard to be innovative: "The touch of anarchy we see in our labs is really a good thing for the advance of physics."

ALVAREZ HAD GONE TO MOSCOW alone, which he later saw as a mistake. "When I arrived home, it was clear to both Geraldine and me that our marriage was failing." The situation had worsened during the dark days of the MTA, which kept him in Livermore while she focused on community service. "Geraldine began spending each summer with her mother and brother at their Lakeside, Michigan, vacation home. For several years she, Walt, and Jean left on the last day of school and returned a couple of days before the fall term started." Alvarez would visit for only two weeks, spending most of his free time in Berkeley on the golf course with Seaborg.

They did what they could to stay together until the children were out of high school—including seeing a marriage counselor—but they didn't quite make it. After Alvarez returned from Russia, they lived under the same roof for another ten months. Using his arrangement with Rowan Gaither, Alvarez bought Geraldine a Ford station wagon, so she could have her own car, and their issues gradually became clear to their neighbors, who until then had been convinced by the show of affection that the Alvarezes displayed in public.

On the evening of March 28, 1957, they finalized the decision. The next day, Alvarez moved out, finding a rented room on Buena Vista Way. "This wasn't a sudden thing, and the children were prepared for it," he told Gaither, but he worried about his relationship with Walt and Jean, as well as the impact on his professional prospects. When Lawrence proposed him as a potential university chancellor, Alvarez was told by Catherine Hearst, a member of the board of regents, that she couldn't approve any candidate who had been divorced.

If they had separated during their rough period ten years earlier, Alvarez admitted, he would have remarried right away, and as it turned out, he wouldn't be alone for long now. On June 12, at an American Society of Mechanical Engineers banquet in San Francisco, he found himself seated next to the vivacious daughter of James Landis, an engineer at Bechtel who served as the group's president. Janet L. Landis was born in Brooklyn in 1930, graduated from Berkeley with highest honors in childhood development, and later studied at MIT. After a stint at the Institute of Child Welfare in Berkeley, she had recently joined the lab to scan emulsions.

At dinner, she peppered him with questions. "My only thoughts of romance that night extended to guessing which of the young men at the laboratory would snatch her up," Alvarez later said. "Some of my friends' wives had sternly warned me not to involve myself with young women. Jan was then twenty-six years old; I would be forty-six the next day." Afterward, he transferred her to his group, which offered more interesting work. "I saw Jan every day at the lab, and a few times at evening parties, but romance never entered my mind—how stupid one can be!"

On the night of August 16, Alvarez was in his office, while Janet checked computer printouts in the next room. William Nierenberg, a professor of physics who was friendly with them both, came over to ask, "Have you seen the comet?" Comet Mrkos, he said, was visible after sunset, complete with a spectacular tail. "We can go up on the roof, and see it across the bay."

They all went up together, but Nierenberg quickly excused himself, sensing that Alvarez and Janet wanted to be alone. In his memoirs, Alvarez fondly recalled the romantic fireworks that followed: "We were discreet about our newly discovered love; for several months we imagined we had fooled everyone. We didn't fool Bill, who arrived home that night from comet viewing and told Edie, his wife, to start looking for our wedding present."

At the end of the year, his divorce was granted, with Geraldine receiving custody of their children. "A divorce then could not be achieved by mutual consent," Alvarez said, "so I agreed to having been 'mentally cruel,' the mildest allowable grounds in California." In the press, it was reported to be on grounds of "extreme cruelty," which was the standard legal language, while Geraldine testified in court that they "argued continuously."

Their son, Walt, said only that it was hard when one's parents "don't get along." Jean, their daughter, remembered Geraldine as a mother who encouraged her children's creativity—she took them to collect jade at a secluded cove in Big Sur—but also as a worthy adversary to Alvarez: "These were two highly intelligent and verbal people with great skills for critiquing others. Let loose on each other, they would have done a lot of damage." Jean concluded, "Having unhappily married parents was terrible—having divorced parents was a relief."

After the divorce, Geraldine, who never remarried, sold the house and moved to Chicago on the day after Walt graduated from high school. In Hyde Park, she lived with Jean and her mother, supporting them as a sales trainer at the Marshall Field department store. Three years later, Jean was diagnosed with non-Hodgkin's lymphoma, a potentially fatal blood cancer. A visit to her in the hospital—she later recovered fully—was the last time that Alvarez ever saw Geraldine.

On December 28, 1958, a week after the divorce decree was finalized, Alvarez and Janet were married at her parents' home in Hillsborough, California. Among the guests was Alvarez's father, who wrote in another context, "It often occurs to me that many a gifted man lives happily for years with a woman who has little understanding of his work and his problems." Alvarez would have disagreed. To one of his students, he offered a simple explanation for the failure of his first marriage: "It was his fault—because he had chosen badly."

For his second wife, he chose a woman who erased the line between his personal and professional lives. At the lab, Janet became a computer operator and later supervised the scanning team. To share in his interests, she took flying lessons and studied physics at the university, with Alvarez signing up to teach her classes. He was pleased that she was two decades his junior—it would help him stay close to the young—and willing to use her intelligence and charm on his behalf, comparing her to the unsung male dancer who lifted up a prima ballerina.

One product of their partnership was a concept for a home fallout meter, which was credited to both Alvarez and Janet. The design—built from common household materials—was based on two foil disks that were charged with static electricity by a comb. It was little more than a Cold War artifact, like duck-and-cover drills and basement shelters, but Alvarez was delighted by the collaboration.

Alvarez later offered a protégé a word of advice: "Many people get married to women that they love, and who are their intellectual equal. Then they develop intellectually and become far more sophisticated—and your wife has not. If you're going to have an experience where you'll grow and mature, make sure to bring your wife along." When he was out of town, he called Janet every night, and he included her in his travels whenever he could.

In July 1959, they went to a conference in Ukraine, where it became obvious that they were under surveillance. One evening, Alvarez noticed that a faulty clasp on his suitcase had been mysteriously fixed, presumably by a Russian agent who assumed that he had broken it while rifling through their luggage. On a train to Kiev, they shared their private compartment with a stranger—apparently from the KGB—who refused to speak or leave. "I amused myself," Alvarez wrote, "by thinking of the mischief I could have caused if I had said softly to Jan that night that Bruno Pontecorvo had asked if I could help him to defect."

Their return flight to Moscow took off from an airfield lined with Tupolev Tu-16 bombers, which a flight attendant warned passengers not to photograph. "It hadn't occurred to me," Alvarez recalled. "My physicist's instincts aroused, I got out my camera. Whenever the stewardess retreated to the rear compartment I snapped a picture." After returning home, Alvarez—who was a trustee of the Mitre Corporation, a systems-engineering think tank for air defense—shared a photo with the Strategic Air Command. The officer there asked, "Where in the *hell* did you get this?"

Janet and Luis Alvarez at his twenty-fifth anniversary party at the Berkeley lab in May 1961. Lawrence Berkeley National Laboratory.

Alvarez, whose Q security clearance gave him access to top secret data, remained on good terms with the military and intelligence establishment. He teamed with McMillan to propose an "electron beam weapon," which Teller funded as the director of Livermore. On a committee known as the Baker Panel, Alvarez had recommended enlarging the National Security Agency. He was especially taken by the codebreaker William Friedman, "one of the most fantastic people I ever met," who had expanded the scale of cryptography as dramatically as Lawrence did in physics.

After Kennedy's election, Alvarez devoted more time to government service. In February 1961, he wrote to Jerome Wiesner, the MIT Rad Lab veteran who headed the President's Science Advisory Committee: "I would like to say explicitly that I am most interested in being of assistance to your office in any way you might wish." He was appointed to an air navigation task force, followed by a committee on limited warfare. As chairman, Alvarez would explore how scientists could contribute to conventional defense, rather than concentrating on nuclear weapons.

The high point was a tour of Asia, including Saigon, where the

United States was providing military advisors and equipment but had yet to commit to a larger presence in Vietnam. Alvarez was casual enough about security to accept an invitation from a relative of Xuong Nguyen-Huu, a Berkeley graduate student, to visit the countryside. Afterward, he learned that two cars of soldiers had discreetly followed, prepared for a firefight if he was kidnapped.

Alvarez felt that his intelligence briefings failed to recognize the resolve of the North Vietnamese, but he drafted a politically neutral report with Daniel Ellsberg, the RAND analyst who later became famous for leaking the Pentagon Papers. Over the summer, he met with the Joint Chiefs and Secretary of Defense Robert McNamara, a kindred spirit who left him "favorably impressed."

Although his work had only a modest impact—Alvarez's primary contribution was a recommendation to develop night-vision technology—he never forgot an informal meeting on August 11, 1961. In Washington, he gave a report to Wiesner, whose office had a view of the Rose Garden. After lunch in the West Wing basement, Wiesner asked if he wanted to see "the boss." Consulting with staff members, they were ushered down a corridor to President Kennedy, who was at his desk, reading newspaper coverage of his administration.

On hearing Alvarez's name, Kennedy asked if he was related to Walter Alvarez, who had once invited him to lunch at home after an examination at the Mayo Clinic. When the conversation turned to affairs in Vietnam, Alvarez—who had seen what could happen to a scientist who spoke his mind—knew exactly what to say. "I shaded the truth when I said that they seemed to be getting better. Most people feel compelled to tell the president what they think he'd like to hear. I was no exception; I'd have had difficulty saying that our policies were disastrous."

THROUGHOUT HIS GOVERNMENT WORK, Alvarez was simultaneously masterminding a seismic shift in experimental physics. At first, it was based on a white lie. He had told Lawrence that the seventy-two-inch bubble chamber was the equivalent of several smaller chambers in a line. In reality, it called for fundamental design changes that he needed to make fast, building up his lead before his rivals even knew that they were in a race. "If we had postponed fabricating the major hardware until we had solved all our problems on paper, the work might still not be in hand."

When the ten-inch chamber was finished in 1956, Alvarez's team installed a target in its own section of the Bevatron, where proton collisions produced a burst of negative kaons. Positioning the bubble chamber to catch it, they were rewarded with a beam for their exclusive use. As the only lab in the world with this setup, they had a "monopoly" over the field for a year, in what Alvarez described as "one of the richest gold mines in the history of particle physics."

The physicist Chien-Shiung Wu—who had left a vivid impression on Alvarez in the thirties in the LeConte basement—had recently demonstrated the existence of parity violation, an unexpected asymmetry in particle decays that he hoped to observe with the kaon beam. Until they had enough experience to train technicians to recognize "interesting events," the physicists scanned all the images themselves, searching for the vertexes that marked an intriguing decay or a direct hit on a proton.

In practice, the beam was contaminated with pions and muons, which at first they ignored. Eventually, however, they began keeping records of an unusual reaction. It occurred when a muon came to a stop, followed by a gap that pointed to an intermediate step involving a neutral particle. When the track came back, the muon was moving again, as though it had been mysteriously rejuvenated. After coming up with a partial explanation, the group turned to Edward Teller, who provided the rest when they visited him at home.

A muon, they concluded, could act like a heavy electron. When it entered the chamber, it was captured by a proton, only to be stolen again immediately by a nearby deuteron. The result was a heavy neutral particle that rebounded away, producing the gap, and then bonded with another proton to make hydrogen deuteride. Because the muon's

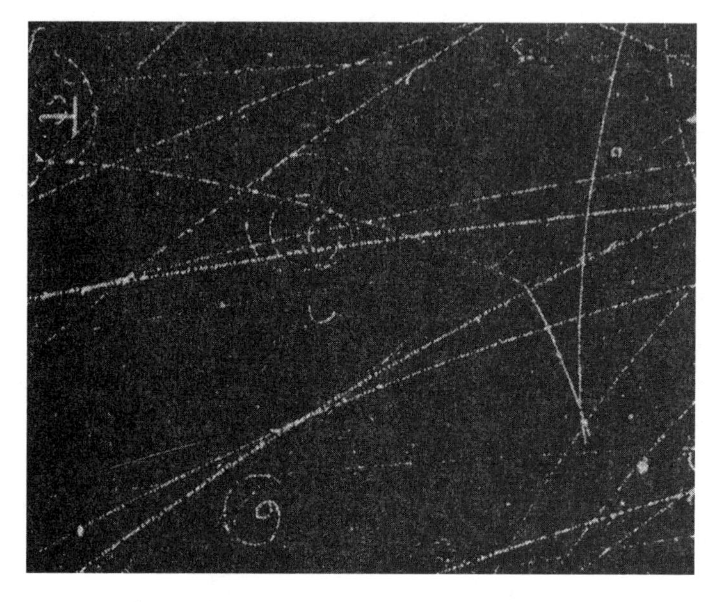

Ten-inch bubble chamber photo of muon-catalyzed fusion.
The American Physical Society.

mass brought the neutron and two protons closer than usual, they fused into helium-3, delivering enough energy to eject the muon once more.

This led to an exciting train of thought. Once the muon was freed, it could catalyze more fusion reactions until it decayed. If they introduced muons into a chamber of hydrogen deuteride, increasing the deuterium supply, it might generate more power than it took to get going. In other words, it was a workable fusion reactor. "For a few exhilarating hours," Alvarez said, "we thought we had solved mankind's energy problems forever."

Years later, Alvarez wistfully recalled, "It would have been the greatest discovery in the history of science." While other physicists were looking to create fusion at millions of degrees, his team had stumbled across a reaction at low temperatures. For now, however, "cold fusion" would remain a dream. After additional analysis, they concluded that the reaction wouldn't break even. "We were off the mark by several orders of magnitude—a 'near miss' in this kind of physics!"

Despite this disappointment, Alvarez was thrilled by what it revealed about the nature of his new tool. Experiments with electronic counters had to be carefully designed for specific objectives: "You are looking for a needle in a haystack." The bubble chamber was entirely different—a visual recorder that captured events that no one had anticipated. "So much other stuff turns up in the course of the experiment that the original problem turns out to be relatively unimportant. It becomes just an initiator for finding out a lot of other things."

Alvarez's messianic sense of the chamber's potential caused tension with colleagues, including Lawrence. Their relationship had already been tested by the Oppenheimer hearing, as the historian Michael Hiltzik noted: "[Lawrence] was gray, tired, and distracted, to the point that he received a rare tongue-lashing from Luis Alvarez." A business manager felt that "Luis was telling Lawrence how to run the lab," and the author Gregg Herken later wrote, "Angered when his own experiments did not receive priority on the Bevatron, Luie hinted darkly about reporting his former mentor to the AEC for misappropriating federal funds."

Lawrence had undoubtedly used laboratory resources on behalf of Chromatic Television, which eventually fell apart, causing his inflammatory bowel disease to flare up from stress. Visiting his hospital room, Alvarez was careful to avoid upsetting him, but the end was closer than either of them knew. On August 27, 1958, Lawrence died after surgery. Alvarez, who served as an usher at his memorial service, believed that his "slow and painful death" was caused by anxiety over his business failures: "The ulcerative colitis which killed him was certainly brought on by those self-destructive thoughts, according to my father, who is an expert in psychosomatic medicine."

It was a horrendous loss—Lawrence was only fifty-seven—that raised the question of succession. On his deathbed, Lawrence had reportedly said to his wife, "Alvarez is brilliant but he's unreliable." Soon afterward, however, the university president, Clark Kerr, invited Alvarez and Janet to dinner. Over coffee in his study, Kerr formally offered him the job. Alvarez was tempted, but he felt committed to the bubble chamber group, which was competing with other teams. After he turned down the directorship of the newly renamed Lawrence Radiation Laboratory, Kerr allegedly responded, "Well, that leaves only Ed."

In any case, Alvarez was preoccupied with the fifteen-inch chamber, which represented a crucial transitional step. Previous chambers had two windows—one for illumination, another to take photos from the other side—which didn't leave room for a rear pole piece that would increase the magnet's strength. Alvarez switched to a single window, despite not knowing how it would work: "I deliberately designed myself into a corner, trusting that I could design my way out." The "coat hanger" reflectors that resulted weren't elegant, but they were good enough.

This became a guiding principle for the new chamber, which was built with a particular goal in mind. Physicists had observed a negative strange particle, the xi, that Gell-Mann's strangeness theory predicted should also come in a neutral version. The trouble was that it promptly decayed into two other neutral particles, neither of which left a track. It might be possible to deduce the parent particle's mass by working backward from subsequent products, but only with a chamber that was long enough for the last step to show up.

In late 1958, it was finally ready. Within a few days, the team found a candidate event for the neutral xi, which it hoped to verify with another. After weeks of searching, however, the bubble chamber broke down. Although he still had just one confirmed image, Alvarez felt secure enough to leave on a honeymoon with Janet—they were visiting Alfred Loomis in Jamaica—on January 24, 1959.

One night, he received a frantic call from a graduate student, Myron "Bud" Good, over a connection that was so bad that Alvarez had to crawl under his bedcovers to shout. He learned that instead of repairing their setup at the accelerator, McMillan had replaced it with a propane chamber developed by another group. Alvarez, who felt a sense of ownership over the beam, later lamented that his absence had enabled McMillan to act "like a tyrant."

It was a decisive move in a battle that had been brewing for over a year. Alvarez had argued that other detector projects should be scrapped as obsolete, while Don Glaser, a recent addition to the faculty, felt that it was best to have multiple approaches. Complaining of Alvarez's dogmatism, Glaser told McMillan, "His behavior put everyone's nerves on edge and could do much to erode the morale and scientific quality of the work at the Rad Lab."

Although McMillan had prevailed for now, Alvarez's team was left with seventy thousand images from the fifteen-inch chamber. While they never found another neutral xi particle, their one confirmed event was solid enough to publish. It was an important result, and it naturally led them to wonder what else they could do with the film on hand, which was all they had for now.

At a weekly meeting, Good proposed that they look for events that produced two pions and the strange particle called the lambda, which were abundant and easy to identify. Once they found the relevant images, they could chart the energies of the pions from each event. If there were any bumps in the data, it meant that some outcomes were more likely than others, in what physicists called a "resonance" that was associated with specific energy levels.

Going through the film, the graduate students Stan Wojcicki and Bill Graziano plotted candidate events as a frequency histogram. Prominent peaks hinted that a resonance was there, but Alvarez saw something even more exciting. Based on his knowledge of the literature, the pattern looked like an interaction with a body of unique mass—in other words, like a particle that no one had seen before.

Alvarez announced it at a conference the following year. The newly discovered particle, the Y^* resonance, could blink into existence when a kaon struck a proton. Before long, his team identified two more—the K^* and Y_0^*—using the same approach. By "bump hunting" in existing film, they were finding particles that theorists had never even imagined.

"It was as if a fisherman with a net designed for catching sharks were catching minnows," Alvarez said. They had expected to use the bubble chamber to study strange particles with much longer lifetimes, making the resonances even more unexpected. Evoking one of his heroes, he wrote, "The search for the particles theorists predict is important, but it's like searching for the source of the Nile or pushing on through hardship to the South Pole; our findings corresponded to James Cook's coming in complete surprise upon the Hawaiian Islands."

As Wojcicki later noted, such happy accidents were really produced by Alvarez's vision of "the ideal way to do particle physics." Initially, however, Alvarez wanted to keep his name—which usually came first alphabetically—off the three papers that resulted, feeling that such a prominent position would overstate his role. The others noted that Margaret Alston, a British physicist who oversaw the analysis, would come before him—not to mention, as her colleague Lina Galtieri later said, that "she was doing all the work." He finally agreed to be listed as an author, a seemingly minor decision that would have enormous consequences for his career.

For Alvarez, it was all just a prelude to the seventy-two-inch chamber, which he sensed would play the same role in his life that the cyclotron had for Lawrence. In 1959, he took a sabbatical year "in residence" to bring it home. As he drew on his contacts and experience from defense work, the author Nuel Pharr Davis wrote, he "tended to treat other scientists like servants." An associate recalled an awkward lunch at the faculty club: "The project business manager came by and Alvarez started chewing him out in front of the whole table. I left and came back and Alvarez was still at it."

The large chamber called for an unprecedented effort at the intersection between cryogenics and data analysis. It was a giant machine—the magnet assembly alone weighed 240 tons—that was walked into place on four hydraulic "feet." Three stereoscopic cameras hung above a horizontal pane of optical glass, five inches thick, that looked into a steel tank of liquid hydrogen. To keep the window from cracking, the temperature was lowered over three days—using compressors from the thermonuclear program on Eniwetok Atoll—to below −400°F.

Cleaning the glass, a technician remembered, was "like looking down the barrel of a cannon." The bubble chamber building—linked by a long pipe to the Bevatron beam—was roofed with plastic panels designed to blow off harmlessly in an explosion, with an emergency

Alvarez at the seventy-two-inch bubble chamber on July 21, 1958. National Archives.

venting system that could dump the hydrogen down a canyon to a spherical pressure tank. Alvarez was still troubled whenever he heard sirens in the night, wondering if they were heading for the lab, and he described the role of a group leader as "the person who talks to the widows after an accident."

None of it would work without his data system, which marked a move into the industrialization of science beyond even Lawrence's dreams. Years earlier, in his blind landing project, Alvarez's "human servos" had been better at seeing patterns than any computer, but only if they had the right equipment. For the bubble chamber, it was a measuring projector called the Franckenstein, named after its inventor, Jack Franck. Its operator, usually a young woman, followed the tracks on the film with a pair of joysticks, using a pedal to stamp the coordinates onto an IBM punch card.

Although the Franckenstein would later be replaced by digital scanners, the underlying premise—the automation of tasks that had once been performed by hand—remained a guiding principle. The raw numbers were fed into a computer that calculated the momentum, angle, and best fit of the particles. Other programs searched for interesting events, printing out summaries for physicists to review, which altered the work itself in ways that not everyone understood.

Alvarez recalled a visit from Melvin Schwartz, a physicist at Columbia, who noticed printouts piling up as the programs were debugged. Schwartz advised him to tell the team to plot the numbers instead: "We'd have the answer to some really important physics in a month or two." Alvarez knew better. With their data-analysis system, the amount of film that they could process would increase exponentially, until they were closing in on a million events per year.

The head count was growing as well. Within a few years, his team would consist of half a dozen faculty, close to fifty graduate students and postdocs, and a hundred scanners. As an independent state within the lab, the Alvarez group developed its own culture, shaped by the strong personalities at the top. Alvarez used a military analogy to explain its dynamics: "If you want to have fighter pilots you have to have guys with fighter pilots' temperaments."

Since the "prima donnas" could no longer be listed on every paper, he looked for other ways to build community. The most significant was a Monday night seminar, modeled on Lawrence's journal club. It was enabled by his marriage to Janet, as well as their move to a larger house on Northampton Avenue. A spacious living room was essential, but so was a wife who could soften his hard edges, making her "a rather major factor in the success of this team effort."

At the weekly meeting, thirty attendees on folding chairs were provided with cookies, pretzels, and beer. It always opened with a summary by a graduate student of recent developments, followed by a speaker whose identity—following Lawrence's precedent—was never revealed in advance. From his armchair in the front row, Alvarez didn't hesitate to ask tough questions. If presenters couldn't answer, he wouldn't let them continue, even if they had barely started talking.

The onlookers were often embarrassed, but the intellectual stakes, as he knew better than anyone, were enormous. Alvarez eventually resigned as an assistant director of the lab, feeling that it conflicted with his responsibilities to the bubble chamber group. It was the second time—after blind landings—that he found himself leading a large team on a project that seemed "so foolhardy in retrospect that it frightens me to think how close we came to disaster."

On March 24, 1959, after four years of development, the seventy-two-inch chamber became operational. Within a few months, it had captured "a sensational picture" of an antiproton beam producing a lambda particle and its oppositely charged equivalent, the antilambda. Two years later, the visiting physicist Bogdan Maglich found a bump in the histograms that he identified as the omega meson, a carrier particle for the strong force, inspiring a congratulatory telegram from President Kennedy.

Given the chamber's expense, Alvarez viewed it through the eyes of an entrepreneur or factory manager. Events were raw materials; citations were products; and the goal was to maximize output per dollar. He focused on variables with a high impact on productivity, especially computer time, and emulated Lawrence by sharing engineering notes with other labs, confident in his insurmountable lead.

At first, Alvarez seemed less willing to share the film itself. Shortly after the large chamber's debut, McMillan met with George Kistiakowsky, who was chairing the President's Science Advisory Committee. In his journal, Kistiakowsky wrote, "Many physicists felt that Alvarez is hugging these data instead of being generous, as any scientist should, and is not providing other groups, without the Bevatron facilities, with opportunities to study the films. Ed conceded that Alvarez was difficult, but assured me that some improvement was being made."

Gradually, he came around, sharing millions of images with other labs, although only after his own group had "skimmed the cream." The film could be analyzed repeatedly to uncover events that had been overlooked the first time. By allowing it to circulate—rather than coming off as "a bad guy" by hoarding it—he increased its perceived value. He was also aware that Seaborg had recommended that he and Glaser receive the Nobel Prize in Physics, and he wanted to build goodwill.

In late 1960, Alvarez received a call from Seaborg, asking if he had heard the news. The Nobel Prize had been awarded to Glaser alone. When Alvarez went downstairs, he remembered, "it was soon obvious from the number of champagne bottles already emptied that I was about the only person in the lab who hadn't heard the news. None of my friends had felt up to the task of informing me."

As Alvarez recalled, "Most everybody in this world wakes up in October of every year, and learns that he didn't win the Nobel Prize, but there are very few of us who have had the unusual experience of knowing that we were really turned down by the jury." Feeling that he had been rejected because of his defense work, he heard later that at least three joint nominations were made, which must have been outweighed by a recommendation to honor only Glaser. "I'll leave it to the reader to guess who might have written such a letter."

Although he had achieved nearly everything he wanted without the Nobel, his confidence was shaken. His need to restore it was part of the reason that he offered his services to Jerome Wiesner, explaining why he had avoided committees before: "I felt that I could do more for the scientific prestige of the country by keeping my nose to the grindstone here in Berkeley." Now that the bubble chamber was "well on the rails," he was ready for a more prominent role.

Away from the lab, his impact was undeniable. Eighteen resonance particles would be discovered in the bubble chamber, either at Berkeley or with the film that it provided. In 1961, Gell-Mann and Yuval Ne'eman independently tried to sort out "the particle zoo" with a revolutionary model known as the eightfold way. The new resonances fit into it perfectly, leading Gell-Mann to propose an even more fundamental building block—the quark. Theoretical and experimental physics were more closely entwined than ever, and it was all thanks to Alvarez's innovations.

His most valued associate was James "Don" Gow, a tremendously gifted engineer who had worked alongside him since the days of the linear accelerator. Although Gow never graduated from college, he served as the chief of staff for the bubble chamber team, as well as the best man at Alvarez and Janet's wedding. After disagreeing with McMillan over a proposed improvement to the Bevatron, however, Gow felt obliged to resign, and Alvarez's contacts offered him the chance to head a nuclear instrumentation firm in Illinois.

Like Lawrence, Gow suffered under the pressure of running a company, for which he had assumed personal financial responsibility. After he went missing on a business trip to Belgium, he was found in a Boston motel, claiming that he had suffered from amnesia for several days. On April 11, 1963, he shot himself in the head at home. His wife found him in the den, a .38-caliber revolver in one hand, with his desk covered in income tax papers. Alvarez was devastated by the death of one of his best friends, blaming it bitterly on "the unfortunate deterioration of Ed McMillan's relationship with our bubble chamber program."

For now, he did what he could to enjoy his outward success. In October, at a presentation of the Collier Trophy to the Mercury astronauts in the Rose Garden, Alvarez was invited onstage with other previous winners. When President Kennedy asked what he had done to receive it, Alvarez recalled, "there was no indication that he knew anything about blind landing of aircraft, but he was gracious." At lunch afterward at their hotel, Alvarez chatted pleasantly with John Glenn.

On November 22, Alvarez—who was scheduled to accept the National Medal of Science from Kennedy later that fall—attended a meeting at Hewlett-Packard headquarters in Palo Alto. Half an hour later, the news came that the president had been shot. While riding in an open limousine in Dallas, Texas, along with Governor John Connally, Kennedy had been struck by at least two bullets, including a fatal injury to the head. Lee Harvey Oswald, the suspect charged in the assassination, was killed two days later in police custody.

Alvarez was horrified by Kennedy's death, which shattered the postwar order that he had sacrificed so much to maintain, and he had no idea yet of the lengths that he would go to seek closure. In Princeton, Oppenheimer learned of the tragedy at the same time. As he left the office, his secretary heard him speak quietly to his son: "Now things are going to come apart very fast."

III.

THE CATASTROPHIST

1963–1988

Richard Feynman was fond of giving the following advice on how to be a genius. You have to keep a dozen of your favorite problems constantly present in your mind, although by and large they will lay in a dormant state. Every time you hear or read a new trick or a new result, test it against each of your twelve problems to see whether it helps. Every once in a while there will be a hit, and people will say: "How did he do it? He must be a genius!"

—Gian-Carlo Rota, *Indiscrete Thoughts*

7.

MONUMENTAL

1963–1970

"I SENSED A STRONG INTERNAL CONFLICT IN LUIE," STAN Wojcicki remembered of his experience in the Alvarez group. "On one hand, he intellectually realized that if high-energy physics was to progress it must be done within the framework of large organizations. On the other hand, he worried that such organizations might stifle originality and provide less than optimal training for students. Down deep, he always remained a gadgeteer and an individualist."

As he surveyed the state of physics, Alvarez was reminded of the period in the thirties when the lab had been content to crank out new isotopes. Now physicists were compiling even more esoteric catalogs. "There is the famous story of the lady who heard an astronomer give a talk and afterward she said, 'Professor, I understand everything about what you do except how you find the names of the stars.' That's what the game is now in nuclear physics, getting the names of the levels."

Because of his own innovations, experimental physics was "just a little dull, so much of the work can be done by technicians." Physicists participated in the engineering only when building the beam for specific experiments; afterward, they simply waited for the printout stage. "A physicist is no longer rewarded for his ability in deciding what histograms he should tediously plot and then examine. He simply tells the computer to plot all histograms of any possible significance, and then flips the pages to see which ones have interesting features."

Since the accelerator operated at all hours, everyone took turns watching meters in the "beam shack" at the Bevatron. During one of his own "owl shifts," Alvarez found that the graduate students there didn't know how the equipment worked—"They had merely been told to push the buttons"—or even where the cables went, since the way was blocked by a "Keep Out" sign. Saying that they would never succeed with that attitude, he gave them a guided tour.

While he took pride in providing "a real educational experience," another physicist felt that the regimented structure of the group itself led to a lack of curiosity: "[Students] ask fewer questions when they spend a lot of time on the beam." The sociologist Gerald M. Swatez wrote that Alvarez's division—which favored experiments that fit into its existing workflow—had a reputation for training its members "to become programmers, not physicists, since the group operates in such a production-line fashion." Alvarez himself admitted, "If I were a graduate student coming into physics now, I wouldn't work in my own group because I would find it too dull."

He had devoted enormous energy to accumulating the prestige that came with a large group, but a managerial role—with junior physicists proposing the actual experiments—seemed unworthy of his talents: "I don't want to spend all my life sitting at a desk reading computer printouts." In a letter to Anne Roe, a psychologist who closely followed his career, he said, "If a person like myself is to keep scientifically productive as he advances in years, he must change his style of doing science." He took comfort in knowing that he had successfully returned to research after the failure of the MTA: "The fact that I made this transition, when almost all of my cohorts fell by the wayside, is one of the things that I am most proud of in my whole life."

Pulling off this transformation again would be even harder because of his government work, which continued to take up half of his time. After the limited-warfare study, Alvarez chaired a military aviation panel, but despite receiving the royal treatment—including a ride in the supersonic Lockheed Starfighter—he was frustrated by his lack of power. He criticized the F-111 aircraft, which was Robert McNamara's pet project, and argued that the C-5 transport would be inferior to a fleet of converted 747s. After both were approved anyway, Alvarez privately described them as "unusually long on cost and severely short on performance."

His next assignment involved the Corona reconnaissance satellites, which monitored missile sites in the Soviet Union. After Francis Gary Powers was shot down over Sverdlovsk, the government moved to replace the U-2 spy plane with satellites that dropped film capsules by parachute. In the fall of 1963, the Stanford physicist Sidney D. Drell assembled a scientific team, including Alvarez, to determine why the images were often fogged. Tracing it to a buildup of static electricity, they became the first civilians to review surveillance photos and secret technical details.

Drell was a product of the Jason think tank, which was established in 1960, under the administration of the Mitre Corporation, to train young scientists to advise on public policy. Five years later, Alvarez became its oldest member. At his first summer session, the Jasons discussed using sensors to target the Ho Chi Minh Trail, which was implemented—with limited success—as Operation Igloo White. "If the current Jasons were to be ranked by their physics knowledge," Alvarez later wrote admiringly, "I would be below average."

On the political front, Alvarez was belatedly awarded the National Medal of Science by Lyndon Johnson, whom he publicly supported over Barry Goldwater. After the 1964 election, however, he scaled back his advisory duties. He told Donald Hornig, Johnson's science advisor, that four years of government committee service "was about as long as a formerly productive scientist could experience, and still recover to become productive again."

Hoping to pivot to private enterprise, Alvarez partnered with the laboratory engineer Pete Schwemin, a "poet of the machine shop," to found an optics company. Its first product—a range finder for golfers—was relatively modest, but other projects were more ambitious. On a vacation in Africa, Janet caught malaria from a mosquito bite, forcing her to spend ten days in bed in Nairobi. To distract himself from his worry over her "terrible chills and fever," Alvarez designed an attachment for stabilizing handheld movie cameras and binoculars.

As Alvarez grew older, he needed to hold books at a distance, until he found one day "that my arms weren't as long as they used to be." He thought that there must be a better solution than the bifocals invented by Benjamin Franklin, but no one had looked into it seriously, since it mostly affected individuals in middle age. "If it were a problem that was of interest to people who are twenty-five years old, I'm sure it would have been done many years ago."

Alvarez came up with a variable focus lens, consisting of two sliding pieces of plastic, that could produce any corrective power. It needed to incorporate exactly the right matching curve, which took him a long time to calculate. Once he found it, he told a graduate student, William Humphrey, who proceeded to derive it independently—to Alvarez's amazement—in less than an hour. Humphrey explained that he had an advantage: "I knew there *was* a solution, and you didn't."

After joining the company, Humphrey invented a diagnostic tool that seemed to work like magic. At an ophthalmologist's office, patients with glasses were often asked to read from a projected eye chart. If the glasses were removed and held at just the right spot in the projector beam, Humphrey realized, a patient would still see the letters as sharp. The result was a device with "phantom lenses" that the doctor could adjust with a knob, with nothing at all in front of the patient's face.

Alvarez's activities were encouraged by Edwin Land, the legendary founder of Polaroid, who had been a key advocate for the Corona satellite program. Land wanted to hire him as a special consultant, but the university refused to release Alvarez from its standard patent agreement, limiting his access to Polaroid's confidential projects. Segrè, he noted acidly, had received an exemption without any trouble—an unwritten benefit of the Nobel Prize.

His rationale for such efforts—aside from "to make a few million dollars"—was perilously close to what Lawrence had said about television. "Business is exciting and challenging," Alvarez wrote, "and as a hobby it certainly beats bridge." Like Lawrence, however, he had trouble deploying his entrepreneurial skills beyond the lab. Despite over a hundred patents, his firm was often close to bankruptcy.

Its most successful invention may have been one that wasn't

publicized. According to Humphrey, Alvarez had many contacts "on the dark side." When an American intelligence agent behind the Iron Curtain needed to take pictures of a document, Alvarez asked Schwemin to disguise a camera as a fountain pen. The gadget—straight out of James Bond—had a strip of film inside, allowing it to be rolled across the page to obtain a clandestine photograph.

In 1966, Alvarez applied his expertise to the most controversial subject imaginable. On the day before Thanksgiving, he had lunch with his graduate students, who were discussing a *Life* cover story featuring John Connally, the second man wounded in the Kennedy assassination. Examining enlargements from the movie taken at the scene by Abraham Zapruder, Connally said that he strongly believed that he and Kennedy were struck by separate bullets.

It was a crucial piece of testimony that cast doubt on the conclusion that Lee Harvey Oswald acted alone. Three cartridge casings were found on the sixth floor of the Texas School Book Depository. One shot missed entirely. Another had been responsible for the president's fatal head wound. This left just one known bullet to account for the remaining seven injuries to Kennedy and Governor Connally.

The Zapruder film confirmed that both men were hit within 1.6 seconds, which was too fast for a gunman to fire twice. As a result, the Warren Commission decided that a single bullet passed through Kennedy's neck, penetrated Connally's back and chest, and shattered the governor's wrist before burying itself in his thigh. Although the "magic bullet" theory was less far-fetched than it sounded, Connally's recollection of two shots clearly required a second shooter.

Alvarez saw it as a chance for the kind of solitary investigation that had become all too rare in recent years. In his study that night, he took a closer look at the magazine, which reprinted twenty-five color stills from the Zapruder film, including the full sequence disputed by Connally. At first, his attention was caught by a flag on the limousine's right fender. Seeing that its left edge was curled in one frame, unlike the surrounding images, he was reminded of the gauges that he had built to measure the force of the atomic bomb, which Fermi had estimated using a few scraps of paper. The flag, he thought, might have been distorted by the shock wave of a gunshot.

When he measured it with calipers, the idea seemed reasonable, although he needed more images to be sure. The Warren Commission had published twenty-six volumes of exhibits and testimony—including all the available Zapruder frames—but most libraries were closed for Thanksgiving. Edwin Huddleson, a Mitre board member and the director of RAND, advised going to the library at San Francisco City Hall. Alvarez spent several hours there, examining maps and diagrams, and later borrowed the relevant volume from the university law school.

He eventually concluded that he had been wrong about the flag, which had only been flapping in the wind—but by now he was hooked. Physics, as he saw it, was more convincing than any witness. It was also a test case of another kind. He felt weighed down at the lab, and by working alone, he could fall back on other advantages. One was his ingenuity; another was his network of contacts, as long as he reached acceptable conclusions. A decade after the UFO panel, he wanted to resolve another debate, but now he was acting as a free agent.

Within a few days, he had focused on a different element of the film. On the limousine's windshield, he saw white spots of reflected sunlight. In one frame, around the time that Kennedy reacted to his neck wound, they turned into parallel streaks, as if the image had blurred when the camera moved. Finding other examples, he hypothesized that they represented moments when Zapruder flinched in response to gunfire. As a source of information on the timing of the shots, it was temptingly straightforward, unlike the morass of eyewitness accounts.

Alvarez wasn't the first investigator to see the "jiggles" in the film, but he analyzed them more quantitatively. From his work on camera stabilizers, he was familiar with neuromuscular jitters. The important part wasn't the streaking, which could appear for all kinds of reasons, but the difference between streak lengths in successive images. The latter reflected the camera's angular acceleration, or how violently it was moving at a given time.

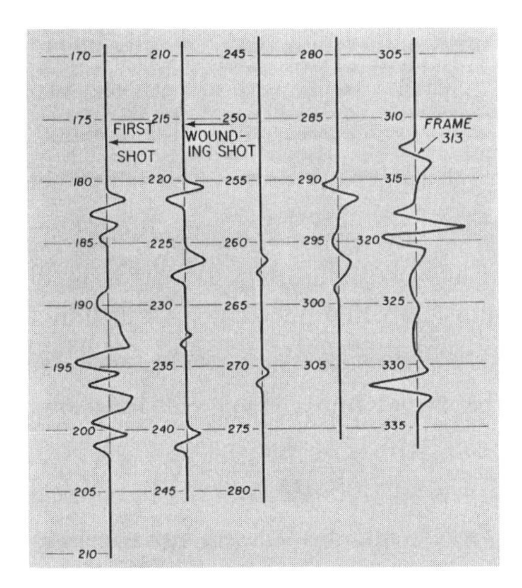

Alvarez's "jiggle analysis" of the Zapruder film, revealing trains of oscillations after frames 177, 215, 285, and 313. The American Journal of Physics.

Using a pencil and tracing paper, he measured the streaks in each image with his calipers. When he plotted their acceleration against the frame numbers, he found four trains of oscillations where the camera was jolted. Assuming that Zapruder would flinch about a quarter of a second—or five frames—after whatever startled him, it pointed to events at frames 177, 215, 285, and 313.

The most interesting movement was associated with frame 177. Since it was considerably before Kennedy's earliest visible reaction, it might have been caused by a missed first shot, which was a potentially vital insight. Based on the film, the two bullets that hit their targets were fired no more than 5.6 seconds apart. If the third wild shot occurred in the same window, the firing rate would have been implausibly close to the limit of what a rifleman could achieve, based on the FBI's conclusion that 2.3 seconds was the minimum gap between gunshots.

On the other hand, if the wild bullet came before the other two—perhaps deflected by the oak tree that briefly obscured the view from the Book Depository—it effectively doubled the time for the shooter to get off one shot, work the bolt action, and fire again. It might also explain Connally's recollection of two separate gunshots, if the first shot that he heard was the one that missed. And the subsequent oscillations after frames 215 and 313 were consistent with the two confirmed hits.

Frame 285 was the problem. It was followed by a similar train of jiggles, although Alvarez later qualified this: "They seemed to me to have less intensity, and to last a much shorter time." All the same, it was undeniably there, and the possibility that it indicated a fourth shot threw everything else into question. Based on the camera speed, no more than 1.5 seconds elapsed between frames 285 and 313, and a lone shooter would have been incapable of firing so quickly.

For now, Alvarez lacked an explanation, which didn't discourage him from spreading the word. On December 4, he sent a summary to Huddleson, who called Richard Salant, the head of CBS News. They also contacted Frank Stanton, the network president and the chairman of RAND. Knowing that a television studio could move more quickly than the government, Alvarez proposed a round of tests for an upcoming program on the assassination. Volunteers were told to hold a film camera steady as a rifle fired at random intervals. As expected, they flinched, leading to streaks that were identical to the ones in the Zapruder frames.

Over Christmas break, Alvarez turned to other problems. One involved the film speed of Zapruder's camera. If it had been set for slow motion, it would shorten the actual time elapsed, ruling out the possibility that one assassin had fired all the shots. Looking for a visual "clock," Alvarez saw a bystander applauding at around four claps per second. With a metronome, he concluded that it would be physically impossible to produce that apparent rate on film if it had been taken in slow motion. The camera, he confirmed, had been running at normal speed.

The other question was how fast the limousine had been moving. Checking the background for stationary reference points, Alvarez noticed a shiny object in the grass, along with people standing with their weight on one leg. Using these benchmarks, he arrived at an average velocity of around twelve miles per hour, falling to eight miles per hour right before the head shot. This was consistent with the mystifying fact that William R. Greer, the Secret Service agent at the wheel, had briefly slowed at a crucial moment, coinciding with the unexplained oscillation after frame 285.

On January 18, 1967, Alvarez examined the original Zapruder film at the National Archives. Five months later, CBS aired a news program, *The Warren Report*, hosted by Walter Cronkite, who was perhaps the only man in the country more trusted than *Life* magazine. In an interview in his living room, Alvarez flatly summarized the significance of his oscillation analysis: "To me it means that there were indeed three shots fired, as the commission said."

There was no mention of the fourth jolt to the camera, for which Alvarez had arrived at what he saw as a plausible cause. In the eyewitness testimony, he found references to "a police siren" that had sounded after the final shot. Feeling free to challenge this point, he later wrote, "The many inconsistencies in the various witnesses' memories made me feel that it was permissible to suggest that the siren, from an escorting police vehicle behind the President's limousine, had sounded a few seconds before the fatal shot, but after the second shot."

If he shifted the timing by two seconds, the siren could explain Zapruder's apparent flinch—and the resulting oscillation—without any need of a fourth gunshot. He also proposed that this was why Greer had slowed down: "When we hear a siren suddenly turned on, just behind our car, we lift our foot from the accelerator pedal." Noting that Greer incorrectly recalled accelerating before the final shot, he used this to underline the fallibility of memory. "That is why I find the photographic record so interesting; it doesn't have the normal human failings."

Alvarez later admitted, "I don't feel as certain about that explanation as I do about the other three cases." In fact, rather than merely discarding "the recollections of some untrained observers," it ignored a single hugely credible witness. Samuel Kinney, the Secret Service agent who drove the convertible behind the limousine, said on the day of the assassination that he started the siren after the final shot: "I observed hair flying from the right side of [Kennedy's] head. With this, simultaneously with the president's car, we stepped on the gas. I released the siren at that time."

At least one critic later pointed out the discrepancy to Alvarez, and his revised timeline was further undermined by the witnesses who reported a third shot, which would have been hard to hear over a siren. There was also a simple explanation for why Greer had slowed. After the second gunshot, he glanced back to see Special Agent Clint Hill racing to climb onto the limousine, and he naturally would have released the gas pedal while looking over his shoulder.

The unavoidable conclusion was that there was no good reason to question the siren's timing. In 1981, Alvarez seemed to admit as much, in a report written while he was serving on another committee related to the assassination: "All witnesses agreed that immediately after the final shot, and as the motorcade speeded up for its dash to Parkland Hospital, sirens were activated." His dismissal of the evidence elsewhere reflected his reluctance to challenge the Warren Report, since any implication of a fourth shot would have destroyed the point of the entire exercise.

In the aftermath, he became the most prominent scientist to endorse the lone gunman theory. Elsewhere, he noted that openly committing to a position was a powerful source of motivation: "I've never seen this factor listed along with necessity as the mother of invention. Edward Teller told everyone for years that he was going to make a hydrogen bomb. . . . Public commitment is often an essential driving force." And it would soon guide many of Alvarez's own choices.

For now, he felt that he had found a reasonable explanation for the fourth oscillation. A more objective observer might have seen two alternatives. One was the involvement of a second gunman, who—for some reason—had fired only one shot. The other possibility, which was perhaps even more troubling, was that physics might not be capable of resolving such mysteries after all. Among the skeptics was one of Abraham Zapruder's relatives, who commented, after hearing about Alvarez's work, "What that guy doesn't know is that all Abe's movies look like that."

❖

SEVERAL YEARS EARLIER, Alvarez had sought a different kind of challenge in Antarctica. It was supposedly motivated by his search for the magnetic monopole, a theoretical particle that might be found near the South Pole, where the structure of the earth's geomagnetic field allowed more cosmic rays to reach the surface. Alvarez easily could have arranged for rock samples to be collected by other scientists, but he wanted to make the trip for its own sake.

His chance came after a conference in New Zealand. Bringing a book about the pyramids, he flew south from Christchurch, arriving after ten hours at McMurdo Station on February 5, 1964. He stayed there for six days, arranging for flights by treating crew members to drinks. On one excursion, while photographing penguins, he was briefly left behind when another researcher boarded his helicopter. "All scientists looked alike to the Navy. When two left and two climbed aboard, the pilot assumed I was one of them and took off."

The visit was more significant for an idea that arose from his reading material. Two years earlier, he had seen the Pyramids of Giza in person for the first time, which naturally led him to wonder how they were built. A hypothesis that the blocks were hauled up earthen ramps struck him as unlikely for simple engineering reasons: "The ramps would have been more massive than the pyramids." He decided that they had been pulled up the side by ropes, using nothing but muscle power.

After returning home, he checked out a stack of books on the pyramids. Studying the cutaway diagrams, he was struck by the differences in their internal structure. The Great Pyramid of Cheops, or Khufu, which was built around 2600 BC, had multiple chambers, shafts, and corridors. Its interior went unexplored for over three thousand years, until a tunnel was hacked into one side. Workmen found the primary chamber, which was displaced from the center to mislead grave robbers, only after hearing the sound of falling stone behind a wall.

The Pyramid of Chephren, or Khafre, was constructed a generation later by Cheops's son, whose likeness was preserved in the Great

Sphinx. In 1818, the explorer Giovanni Belzoni discovered an empty burial chamber cut out of the underlying bedrock. Apart from its two entrances, there were no known cavities in the pyramid itself. The consensus among Egyptologists was that the rooms in earlier pyramids reflected a period of experimentation that had ended by Chephren's time.

Alvarez was unconvinced. Rejecting the conventional wisdom of experts, he asked himself, "Why would Chephren, after a boyhood spent watching his father's slaves erecting a beautiful and complex series of chambers and passages in the Great Pyramid, be content to erect a solid and uninteresting pile of limestone blocks as his own pyramid?" Although a pharaoh might well have chosen more visible displays of splendor, the question felt like enough to justify further investigation. "If Chephren's grandfather built two chambers and his father three, it seemed most likely to me that Chephren would have ordered four."

Obtaining permission would be a delicate process, which was part of why Alvarez was drawn to it. He wanted to develop lines of influence outside the lab, and the pyramid search, with its aura of romance, was an ideal place to start. To look inside, he devised an ingenious method. In 1955, an Australian hydroelectric project had used Geiger counters to estimate the mass of rock above a tunnel, based on the radiation that made it through from space. The Dutch physicist Samuel Goudsmit had proposed a similar approach for the pyramids, and Alvarez refined it to focus on muons, which were abundantly created in atmospheric collisions with cosmic rays.

On March 1, 1965, Alvarez finished a proposal for "X-raying" the Pyramid of Chephren. As an analogy, he noted that it would be easy to construct a pyramid model, put a radiation source inside, and capture an image of the interior on film. With the actual pyramid, they could do the same in reverse. Muons in cosmic rays lost energy as they passed through matter, "just as a rifle bullet does when it bores its way into a fence post." If the amount of stone in the pyramid was decreased anywhere by an intervening cavity, more particles than usual could be counted on the other side.

In the Belzoni chamber underneath the pyramid, Alvarez proposed installing a set of spark chambers, each consisting of two metal plates about a centimeter apart. When a muon crossed the gap, it produced an arc of electricity along its path. By analyzing the direction of hundreds of thousands of events, a computer could produce a map of the interior, like a meteorological chart. If a region of low density was detected, the spark chambers would be relocated, allowing the exact location to be determined with simple trigonometry.

Alvarez was confident in the basic idea, but he needed financial and political support. He obtained a referral to Fathy El Bedewi, a physicist at Ain Shams University in Cairo, and looked for a prominent Egyptologist. Because he admired the book *The Pyramids* by Ahmed Fakhry, Alvarez arranged to meet with its author, who agreed to participate. For the rest, he turned to his contacts. Hewlett-Packard furnished electronic equipment, and he secured funding from the AEC, the Smithsonian, and the National Geographic Society.

An agreement between the United States and Egypt was announced on June 14, 1966. Bedewi, Fakhry, and Alvarez would head the project, at an estimated budget of $250,000, while Jerry Anderson of the Lawrence Laboratory would run the team. Three months later, Alvarez paid a short visit to Egypt with Janet and Walt to review their progress. Most of the hardware was installed a quarter of a mile from the pyramid, with cables running to the Belzoni chamber. An IBM 1130 computer in Cairo would receive the magnetic tapes, followed by analysis by Alvarez's group in Berkeley.

On February 18, 1967, Oppenheimer died of cancer in Princeton, where a service in his memory was held the following week. Deciding that it was best to stay away, Alvarez spent the morning on the golf course. As a result, he missed the eulogy by Henry DeWolf Smyth, the only AEC commissioner to vote to renew Oppenheimer's clearance, who told the six hundred attendees, "Such a wrong can never be righted; such a blot on our history never erased."

A month later, the American team arrived in Egypt. Although the program was seen as a model of international cooperation, conflicts

Alvarez and associates with the spark chambers to be installed inside the Pyramid of Chephren. Left to right: Jerry Anderson, Fathy El Bedewi, Alvarez, James Burkhard, Lauren Yazolino, Ahmed Fakhry. Lawrence Berkeley National Laboratory.

arose over the chain of command. Anderson insisted that he answered only to Alvarez, while Bedewi informed the visitors that they needed his permission to enter the lab, to which he held the keys. A walkout by the Americans halted work for three days, and it was barely resolved before a brief visit by Alvarez in the spring.

On June 4, the spark chambers successfully operated for the first time. The following day, after months of rising tension, a surprise Israeli assault destroyed most of the Egyptian air force. In the wake of the Six-Day War, Israel took possession of the West Bank and Gaza Strip, straining relations between Egypt and the United States, which was suspected of aiding the attack.

The pyramid team was in Cairo when the conflict erupted. When the police saw Jerry Anderson filming a home movie of the crowds, he claimed at first to be a Danish citizen. The Americans were put on a train to Alexandria, but on their arrival, Anderson was separated from the others by armed officers and forced into a car. Under interrogation, he was reportedly threatened with a pistol, made to stand at attention with his face against a wall, and called "a dirty spying American dog." Although he managed to talk his way out, the team hushed up the incident.

Sensing that the Egyptians were suspicious of the Americans, Alvarez indulged in "an interesting fantasy." He imagined that Egypt had learned about the equipment in the pyramid—"the finest of bomb shelters"—and his background in radar, which implied that his group had somehow enabled the Israeli planes to elude detection. As they waited for tensions to ease, Fakhry stationed a guard over the spark chambers, and work resumed only in 1968. "We were welcomed with open arms," Alvarez wrote, "indicating that my fantasy was just that."

At the end of March, Alvarez spent a week checking up on their progress in Egypt, where Lauren Yazolino had taken over from Anderson. The spark chambers were placed horizontally at first, exposing them to a vertical cone—about a fifth of the pyramid's volume—that seemed like the most probable location for a room. As the team began generating histograms of the muon counts, Yazolino wrote on September 21 that he was excited by the results: "I think you had better plan to come to Cairo. We are almost there, Luie."

The following day, Alvarez received an electrifying telex in Berkeley from Yazolino: "There is no doubt we have found an EGG." As soon as he realized that this stood for "East Grand Gallery," he booked a flight to Egypt. On the plane, he dreamed excitedly of what they might find when they drilled through the wall. The treasures in an undiscovered chamber would surpass the tomb of Tutankhamun, making it "the greatest discovery in the history of archaeology."

At the university, Alvarez eagerly examined the histograms, which showed a region where the muon count was twice as great as average. As the champagne flowed, however, Yazolino saw that Alvarez seemed hesitant. After spending the night thinking it over, he returned the next morning, convinced that something was wrong: "A chamber that big would certainly cause the pyramid to collapse."

Before long, they found an error in the code. The counts had been sorted into "bins" corresponding to different angles, which were then associated with areas of the pyramid itself. With a sinking sensation, they saw that the central region had accidentally been assigned muons from more than one angle, doubling the apparent intensity. The chamber was a mirage.

It was an embarrassing mistake, but Alvarez claimed not to mind: "If they had found it themselves, I would have missed two of the most exciting days of my life." Elsewhere, he explained his reaction in technical terms:

> In addition to the real things that one finds, one "discovers" many things that turn out on a more careful examination to be false. If human beings were "linear systems," the disappointment on learning that a discovery was false would just cancel out the elation built up when one believed the discovery to be real. But fortunately, we are not linear systems, so the overall time-integral of excitement and pleasure is strongly positive—the enormous area under the elation vs. time curve is hardly diminished by the small negative area in the letdown phase. For me this is a very powerful effect.

He treated this as a universal aspect of human nature, but it was clearly more true for him than it was for most people. It explained his unusual tolerance for failure and risk, as well as his impatience with colleagues who lacked it—and his own compulsion to keep chasing that feeling of excitement.

ONCE THE DEFECTIVE CODE was fixed, the team resumed its work in Egypt, gathering thousands of counts over the months that followed. As the tapes were fed into the computer to generate a scatter plot, a shadowy image of the pyramid appeared, like a slowly developing photograph. Finding one chamber would justify the entire investment, vastly increasing the resources at Alvarez's disposal in the future—but only if he emerged with a success that the world could understand.

Alvarez was already at a personal crossroads. He was about to turn fifty-seven, an age when many men would be content to take satisfaction in their grown children. Walt had moved to Libya with his wife, Milly, to work in petroleum exploration, while Jean—who had attended Oberlin and Purdue—was a physical education instructor at Wellesley. Instead of settling into his golden years, however, he wanted to be a father again. In 1965, Janet gave birth to a son, Donald, named after the late Don Gow, and a daughter, Helen, followed two years later.

He was less happy at the lab, where he had sensed a change after McMillan took on the directorship. Alvarez claimed that Clark Kerr, the university president, had expressed doubts about the choice at the time. "I agreed that Ed's life history suggested none of the administrative skills or interests that one would want in a director, yet I said he was the only really distinguished scientist available." In the end, Kerr allegedly decided to appoint McMillan for a trial period of five years.

According to Alvarez, the shift in atmosphere was evident at once. In the past, his colleagues had often used him as an intermediary to bring issues to the director's attention. One day, he visited McMillan in Lawrence's old office, hoping to have an informal talk about a minor problem. "I assumed, without any conscious thought, that I would act with my longtime friend Ed as he and I had with Ernest. But Ed let me know that I was overstepping the bounds of propriety."

Alvarez knew better than to try again. For a time, he thought that the decline in their relationship was because of his divorce—Geraldine had been close friends with McMillan's wife, Elsie, who also happened to be Molly Lawrence's sister. Eventually, he conceded that Elsie had been perfectly welcoming to Janet, whom she invited to join the laboratory wives at their monthly dinner, and the conflict was really over who would control the lab.

The result, he felt, was a series of grievous errors: "Every high-energy physics decision that Ed made in this period was in my opinion a serious mistake." In response, Alvarez sought outside funding and consolidated power within his own circle, which left him feeling even more confined. As he competed for resources, he became convinced that McMillan was deliberately undermining his group in favor of other teams, which only made them all "equally weak."

In 1963, McMillan's reappointment led to what Alvarez recalled as a rare agreement with a rival. "I had a visit from a disturbed Emilio Segrè. Emilio said that neither he nor Edward Teller had been consulted and if asked, they would have advised against it. I was of the same persuasion. We all felt that another five years of Ed as director would be a disaster that the laboratory might not survive." In his memoirs, Segrè was more forgiving of McMillan, who had to deal with "the implacable personal hostility of Alvarez, who antagonized him constantly."

After McMillan's position was secured, he transferred technicians to competing programs for measuring bubble chamber film, which Alvarez called "total flops." Their most hostile exchange occurred when Janet was offered a research position by the biochemist Melvin Calvin. McMillan refused to allow it, invoking a rule against nepotism, which Alvarez argued had been waived for others. Noting that Janet didn't need the money, he told McMillan, "I do feel, however, that she should not suffer because of your personal feelings toward me."

As the tensions became impossible to ignore, Alfred Loomis proposed an unusual solution. At a meeting in his apartment, which was also attended by Seaborg, Loomis offered to pay for the Alvarezes to travel for a year "so that Ed could get his act together." Neither McMillan nor Alvarez found the plan acceptable, so it was quietly discarded. In the aftermath, Alvarez decided that there was just one way out. As the Kennedy investigation, the pyramids, and his other projects had demonstrated, there was a world elsewhere.

On March 18, 1968, Alvarez wrote to McMillan that he had fulfilled Lawrence's request that he remain with the bubble chamber until it was firmly established. Since then, the program had "passed its prime"—the chamber itself had gone to Stanford—and he pointedly noted that nothing had taken its place. "In my opinion, this simply means that experimental high energy physics is not as interesting a field now as it was a few years ago." McMillan, in turn, had failed to support studies of cosmic rays, where exciting research remained to be done: "Certainly Ernest Lawrence would never have adopted such a course of action in his laboratory."

Alvarez coldly summarized his predicament. He wanted to run experiments on cosmic rays with backing from NASA, but McMillan discouraged him from seeking support from other agencies, while also ruling that the subject was too far from the lab's mandate to be funded by the Atomic Energy Commission. Since he lacked the discretion to pursue the ideas that he found worthwhile, Alvarez said that he had no choice but to resign as a group leader.

The consequences were immediate. Alvarez wrote that he would cancel his Monday night seminars—a position that he later reversed—and asked only to keep his office and parking space. The Alvarez group, renamed "Group A," unanimously elected Frank Solmitz, its head of data analysis, to replace him, which testified to the importance of computers in its work. Setting up shop in another building, Alvarez focused on cosmic rays, while impatiently awaiting the results from Egypt.

On October 29, at a party hosted by Melvin Calvin, Alvarez met the wife of David Perlman, the science editor of the *San Francisco Chronicle*, who remarked, "Oh, you're one of the Berkeley Nobel laureates." Alvarez responded—as he often did—that while many of his colleagues, including Calvin, had won the Nobel Prize, he had yet to receive it.

Alvarez was scheduled to fly out the next day for a conference on cosmic rays at the Manned Spacecraft Center in Houston, so he and Janet returned to their new house on Southampton Avenue. At 3:15 in the morning, he was awakened by the telephone in the bedroom. When Alvarez answered, he heard a reporter from CBS in New York informing him of an announcement that had just come over the wire. He had won the Nobel Prize in Physics.

It was an honor that he had desperately wanted eight years earlier, only to conclude that he had missed his chance—and the long wait left him even more prepared to take advantage of it. Outside the lab, at least, it would put him on equal terms with McMillan, with a source of power that he could use in countless ways. All of this would occur to him soon, but as he sat in the dark, processing the news, he managed only to say, "You're kidding."

WHEN JANET WOKE UP, Alvarez told her what had happened. Events would soon move very quickly, and he wanted to confer with his wife—his most trusted confidant—while he still had the chance. The prize would open many doors, but there were also pitfalls to avoid if he wanted to fully utilize it in the time that he had left. "I made two resolutions. I wouldn't sign petitions, and I wouldn't accept social invitations that were tendered only because of my Nobel Prize."

Wondering why no one else had called for the last ten minutes, he saw that he had left the phone off the hook. Once it was back on the handset, it began ringing with congratulatory calls. He and Janet got dressed, knowing that photographers would soon be on their way. Alvarez went downstairs to check whether the story was in the paper—it was—and headed across the street to the house of his colleague Arthur H. Rosenfeld. Hearing the doorbell, Rosenfeld peeked out his bathroom window to see Alvarez, who eagerly asked, "Have you seen the *Chronicle*?"

When the official congratulatory telegram arrived from Sweden, it stated that Alvarez had been recognized "for your decisive contributions to elementary particle physics, in particular the discovery of a large number of resonance states, made possible through your development of the technique of using hydrogen bubble chamber and data analysis." He noticed with satisfaction that it was the longest citation in the history of the prize. "I don't attribute any significance to this, but my job is observing, and this is a fact that I observed."

He later reflected on the role of chance. The resonances mentioned in the citation had been announced in three articles that he originally hadn't planned to sign. He changed his mind only after confirming that he would appear second, as an author, after Margaret Alston—and if he hadn't been listed, he might never have won the Nobel. "A bookie who counted pages in the phone book would say that for

Margaret to have a last name ahead of mine alphabetically would be a 42-to-1 longshot."

According to the secretary of the Royal Swedish Academy of Sciences, the decision "unlike many other times was absolutely clear." Combined with computers, the bubble chamber had vastly transcended Glaser's initial conception. Alvarez learned that Seaborg had pushed hard on his behalf, and he was glad to have finally fulfilled the hopes of his colleagues, including Lawrence: "Ernest had told his brother John on several occasions that he felt sure that I would eventually win it."

After canceling his flight to Houston, Alvarez drove to the office, which was decorated with balloons labeled with the resonance particles that his group had found. McMillan, he noticed, was conveniently away in Pasadena, but the director held a party at the lab the next day. He also loaned Alvarez a white vest that had been worn to the ceremony by most of the Berkeley winners.

In December, Alvarez flew to Stockholm, accompanied by Janet; Walt and his wife, Milly; and eight group members and their spouses, whose tickets he bought with his prize money. At the banquet, a student delegation raised the problem of population growth. In response, Alvarez joked that he was surprised "to see no men in long hair and beards, and to see no girls in secondhand Army shirts and patched trousers." He noted that there were "no simple solutions" to social issues: "Physics is the simplest of all the sciences; it only seems difficult because physicists talk to each other in a language that most people don't understand—the language of mathematics."

A physicist, Alvarez continued, could predict changes in an ideal system, but it was harder in human affairs. As an example, he noted that caesarean sections allowed babies with larger heads to survive, which might lead to higher mortality in later generations if medical care was disrupted by "an ice age, a period of intense volcanic activity that could blanket the atmosphere in dust, or a melting of the Antarctic ice cap that could flood most of the land." Revisiting his remarks long afterward, he was struck by how catastrophes were already on his mind.

The following afternoon, he delivered his Nobel lecture, expressing gratitude that he had been recognized for both his discoveries and his inventions: "If we spend all our time developing equipment, we risk the appellation of 'plumber,' and if we merely use the tools developed by others, we risk the censure of our peers for being parasitic." He warmly thanked Stan Wojcicki and Bill Graziano, whom he later described as "my Jocelyn Bells"—a nod to the astrophysicist who had identified the first pulsar but saw the Nobel Prize awarded to others.

There were also notable omissions. Years earlier, in the dispute over accelerator upgrades, McMillan had sided with Edward J. Lofgren, the director of the Bevatron, which Alvarez saw as a "blatant exercise in pure cussedness" that led to Don Gow's resignation. In his speech, he didn't mention McMillan, and he later stated, "I purposely refrained from thanking Ed Lofgren and the people who ran the Bevatron because I knew of the serious roadblocks they had placed in our way."

A few years later, a newspaper reporter wrote, "Students and colleagues at UC Berkeley say physicist Luis W. Alvarez was testy before he won his 1968 prize, but mellowed considerably afterward." In private, his ambition was undiminished, and he often put his mind to obtaining honors for his colleagues. He largely lived up to his pledge to avoid petitions, with one notable exception—he ultimately joined over forty Nobel laureates to call for "swift disengagement" from Vietnam.

Given the advantages of his heightened profile, he planned to scale back his involvement with large groups, but he still had unfinished business. The computer analysis of the Pyramid of Chephren had produced a number of "very exciting signals," one of which seemed to show a chamber directly under the peak. Early the following year, he told Seaborg, "If it refuses to go away for another few weeks, we'll have an exciting announcement to make in Washington."

In April 1969, at the American Physical Society meeting in Washington, DC, Alvarez—who was serving as the group's president—finally revealed the results. After counting over a million muons, they found that any apparent cavities simply melted away as they refined the program to correspond with the pyramid's geometry. The central area of the pyramid was completely solid.

This outcome was widely characterized as a "failure," which infuriated Alvarez. Although he insisted that the idea had worked perfectly, he was unable to conceal his disappointment: "I have never been so surprised at the results of any scientific experiment in my life." He also pledged to search the remaining eighty percent of the pyramid. "If we pull out now, Chephren will win."

Seeing it as a test of his newfound influence, he scraped together the funds from the National Science Foundation, and another round of muon counts began three years later. One tantalizing void lingered for weeks, but it disappeared when they pointed the counters in its direction. Like the rest, it was a statistical fluctuation. The Pyramid of Chephren was nothing but a huge pile of blocks.

Whenever anyone said afterward that he didn't find a chamber, Alvarez never hesitated to set them straight: "We found that there wasn't any chamber." He was also annoyed by the "pyramidiots" who saw a mystical significance in the results. Many cited a statement by a computer scientist at Ain Shams: "There is some force that defies the laws of science at work in the pyramid." The quote evidently referred to the brief period when a coding error pointed to an impossibly huge chamber, but it was widely reprinted out of context—and often misattributed to Alvarez himself.

In fact, Alvarez eagerly debunked subjects like telepathy, psychokinesis, and "pyramid power," which implied that a pyramidal container could restore a razor's sharpness, and he was delighted when his colleague Sharon "Buck" Buckingham found no change in a blade that was left in the Great Pyramid itself. His work on UFOs and the Kennedy assassination led to many crank letters, which he kept in his "nut file." Strangers would occasionally come uninvited to his house, and his children learned to hang up on any caller who claimed to be a distant relative. He always replied politely to his correspondents, however, partly to avoid provoking a violent reaction.

Alvarez was also tactful with William Shockley, who received a Nobel Prize for his role in the invention of the transistor. Although they had served together on the limited-warfare committee, he was wary of Shockley's racist views on eugenics. When Shockley tried to enlist his support, Alvarez thanked him for the "interesting material," but he declined to get involved. Elsewhere, he described Shockley as someone who "comes on pretty strong in an area in which he is not an expert," seeing him as a cautionary tale of a scientist who damaged his reputation.

He became even more careful as he sought funding from a potentially generous new source—the space program. Ever since Sputnik, he had seen it as both a competitor for government dollars and a benefactor that he could cultivate independently of McMillan. By the sixties, he was ready to take advantage of it, noting that the public was turning away from less glamorous subjects: "You have to get an appeal into the science of physics or the expenditures will go into space."

In June 1963, Alvarez had attended a space sciences conference in Warrenton, Virginia, organized by James Webb, the administrator of NASA. One result was the establishment of a physics committee consisting of Alvarez, Murray Gell-Mann, Willard Libby, and other prominent researchers. It was led by Homer Newell, the associate administrator for space science and applications, who Alvarez knew "had extensive discretionary funds under his control."

What happened next would be described by the scholar Peter Galison as "one of the great twists of physics history." After spending close to a decade seeking to replace cosmic ray research with the bubble chamber, Alvarez made a decisive pivot back to a field that had fascinated him ever since the Chicago World's Fair. He wanted to study cosmic rays with balloons.

On May 4, 1964, Alvarez finished a remarkable proposal. Until about ten years ago, he wrote, all research on unstable particles had been conducted using cosmic rays. More recently, large accelerators had taken the lead, but they were unlikely to reach beam energies of one hundred billion electron volts, which put a hard limit on the production of exotic particles in the laboratory.

Yet there were plenty of protons at that energy level available right now—in cosmic rays. Accelerated by shock waves from distant supernovas, they were more powerful than could be achieved in any lab. As long as the target was made as large as possible, Alvarez argued, the rate of measured events from cosmic ray bombardment would be comparable to the accelerator and bubble chamber.

To capture protons before they interacted with particles in the atmosphere, the experiments had to occur at extreme elevations. Even a mountaintop wouldn't be high enough to register the necessary number of events or avoid contamination. Doing it properly, Alvarez said, required altitudes of up to one hundred thousand feet, which were achievable with commercial balloon services: "We shall therefore assume that we do not have to become balloon experts."

The equipment itself amounted to an automated physics laboratory in a gondola. First came a Cherenkov detector, a counter that picked up the shock wave of light produced by a proton in the "interesting" range, which would then pass through two distinct setups. One was designed to study cosmic rays themselves, while another would use protons as the equivalent of an accelerator beam on a liquid hydrogen target. To work independently of the power grid, it needed a superconducting magnet that functioned only at cryogenic temperatures—the formerly "missing element," now technologically feasible, that made the rest possible.

Even by Alvarez's standards, the proposal was audacious. It called for a combination of technologies as ambitious as the bubble chamber; it had to operate without human intervention for twelve hours at a time at a height of eighteen miles; and he was asking for a sizable $3.8 million. Despite McMillan's refusal to allocate money to cosmic rays, Alvarez eventually received enough from NASA to cover the project and pay his own salary. It gave him the confidence to resign as the leader of the bubble chamber group, which he left to establish his own operation in Building 46.

As he found himself serving on advisory committees on space, Alvarez also attracted fresh talent. "Young graduate students that I think are the best available tell me . . . that they wouldn't work in my group if it weren't for the balloons. They find this very exciting, to get their hands on some equipment. They don't spend all their time looking at printout. So I get a lot of personal satisfaction out of perhaps being able to start turning the wheel around in the other direction."

In 1965, NASA offered him the directorship of Goddard Space Flight Center, which he declined in favor of continuing his High Altitude Particle Physics Experiment. A year later, his group conducted a test flight in Palestine, Texas. Although it succeeded, the unsteerable balloon could be launched safely only during "turnaround" periods in the summer and autumn, when the jet stream decreased in speed. To increase the available window, the project switched to a site where a balloon would drift harmlessly over the ocean. As a side benefit, Alvarez bought a Cessna 310 monoplane—a Christmas gift to himself—to commute to the new headquarters in Chico, California.

At five in the morning on August 9, 1967, they launched a trial balloon. From a telemetry station in Mendocino County, Richard Muller, the graduate student who had proposed the test, monitored it by radio as it headed west over the mountains. Nine hours after launch, it was a hundred miles out to sea. A recovery barge transmitted a signal to fire a squib that would release the gondola, but the payload vanished without a trace. "We never found the balloon, parachute, or science package," Muller said. "We guessed that the parachute failed to open and that my apparatus slammed into the ocean at more than 100 miles an hour."

The loss had a chilling effect on the project, which was also set back by the founding of the National Accelerator Laboratory in Illinois, later known as Fermilab, where energies would be produced as great as those in cosmic rays. Faced with wavering support, Alvarez streamlined the program. Rather than a combination of astrophysics and accelerator physics, it would focus solely on the former.

Balloon work was still frustrating, and after the Chico "fiascoes," they returned to Texas for a series of launches starting in July 1970. Joining the team was George Smoot, a recent arrival from MIT, who recalled Alvarez as "a leader by intellect, example, and fear." As Smoot fought off bugs and scorpions, the balloon project challenged him as nothing else ever had. "It also, in the end, broke my heart—or at least dampened my enthusiasm for this brand of scientific adventure."

Their first experiment was inspired by Hannes Alfvén, a Swedish physicist who proposed the existence of entire galaxies of antimatter, set apart from our area of space by a boundary region formed by the mutual annihilation of particles and antiparticles. From a distance, it was impossible to visually distinguish an ordinary star from an antistar. Cosmic rays, by contrast, which traveled vast distances, offered a physical sample of material from elsewhere in the universe.

Alvarez decided to search the cosmic rays for proof of Alfvén's antimatter galaxies. If they contained heavy nuclei of antimatter that could be forged only in antistars, such as antioxygen, the particles would curve in the opposite of the usual direction in a spark chamber. The physicist Andrew Buffington failed to find any such events in the first round of balloon data, but Alvarez asked to see all the candidates. Given the huge potential upside, he devoted unusual care to examining this "enriched sample" of four dozen tracks, out of the tens of thousands that were captured.

His attention was caught by a single provocative image. The setup in the balloon included three scintillation counters that a particle would trigger in succession. Alvarez noticed that one event deposited the most energy at the bottom, or the reverse of what was normally seen. Given the possibility that it was caused by an antioxygen nucleus annihilated in an impact with an ordinary atom—which would produce a large pulse—they took another look at the film, where they found a negatively curved path consistent with antimatter.

Before going public with such a revolutionary claim, they had to rule out all other explanations. The apparent curvature, for example, could have been produced by alignment errors in the viewing mirrors, while the unusual pulse might reflect a hard collision by a commonplace particle. In the end, they calculated that it was three times more likely to be antioxygen—which wasn't enough. To meet the threshold for publication, the odds had to be a million to one in its favor.

Although the outcome was disappointing, Buffington and Smoot were struck by Alvarez's attitude. When a promising candidate was found, he insisted on making absolutely sure of its significance, neither dismissing it without an investigation nor embracing it prematurely. He was determined to overlook nothing, and he often invoked a rule that would guide his actions in the dramatic decades to come: "Let's be sure to give this event a decent burial."

8.

HOLD EVERYTHING

1969–1982

IN 1969, ALVAREZ'S ATTENTION TURNED TO THE MOON. SIX years earlier, at the NASA scientific conference in Virginia, he had spoken with the physicists Lyman Spitzer and Bruno Rossi about a paradox in astrophysics. If cosmic rays were ejected by supernovas, there wasn't enough room in our galaxy for them to accelerate to such high velocities. Their enormous energies hinted at an extragalactic source, but they were more uniform than would be expected if their origins were widely dispersed.

On returning to Berkeley, Alvarez happened to see a preprint of an article on magnetic monopoles, the hypothesized particles that had fascinated him for years. He felt a sudden collision between two seemingly unrelated ideas, which was how insights often arose: "With the galactic and cosmic ray numbers so recently reinforced 'in the memory,' it was natural to inquire if the existence of monopoles would eliminate the paradox of the high energy cosmic rays."

The magnetic monopole was an intriguing target. An ordinary bar magnet had two poles, conventionally labeled "north" and "south." If you broke it in half, you didn't end up with separate north and south poles but with a pair of new magnets. This held true down to the subatomic level. Poles were never seen in isolation, even though Maxwell's equations indicated that electrical and magnetic forces ought to be symmetrical. The glaring exception was the existence of particles with a net electric charge, like the proton and electron, and the apparent lack of any magnetic equivalent.

In 1931, a paper by the English theorist Paul Dirac reignited interest in the possibility. Dirac envisioned an interaction between a single electron and a hypothetical monopole, which respectively had the smallest imaginable electric and magnetic charges. His analysis implied that the product of these charges—which produced an infinitesimal electromagnetic field—was an integer multiple of a specific unit, or what was called a quantized property.

He drew a grand conclusion. If just one magnetic monopole existed in the entire universe, electrical charges would always be quantized. Unlike mass, they couldn't be any arbitrary number, but a multiple of the charge of one electron. Since this had been confirmed by countless observations, physicists naturally wondered if the monopole might also exist. Dirac contented himself with a dry understatement: "One would be surprised if nature had made no use of it."

Not surprisingly, Alvarez was drawn to the problem, which would restore symmetry to classical electromagnetism. It occurred to him now that it might also explain how cosmic rays could attain immense energies within our own galaxy. If magnetic monopoles were produced when the rays collided with hydrogen in interstellar space, they could bind with protons, creating a package that was accelerated by magnetically charged galactic clouds.

Alvarez thought that the idea was enough to justify a renewed monopole search. Even if the odds of success were remote, the upside was huge, and it offered another opening for a push into astrophysics. Most astrophysicists were trained in observation, not experimentation, which gave him a distinct advantage. Since monopoles had never been seen in a lab, they were probably too massive to be produced by accelerators, so he had to invent a way to find them elsewhere.

His first thought was to check locations exposed to cosmic rays, which could deposit monopoles from space or make new ones in collisions. Most places on the earth's surface had been churned by geological activity, making it unlikely that anything would be found there. A potential exception was near the South Pole, where the geomagnetic field guided cosmic rays to the ground. Other possible sources included meteorites and the slowly deposited sediment on the ocean floor.

To search the samples, Alvarez designed a machine based on

the principle that a current would be induced by a monopole passing through a coil of wire. Pellets of pulverized material were placed around the rim of a metal wheel, which spun at a hundred times per second through a pair of coils. If a monopole appeared, a stylus that recorded the electrical output would show a telltale spike.

A technician named Lincoln Kilian was told to check rocks that Alvarez had collected in Antarctica, meteorite fragments, seafloor cores, and elements from the laboratory storeroom. One morning, he saw the pen jump, prompting Alvarez and his colleagues to rush over. "All were smiling nervously," Kilian wrote, "and the air nearly crackled with anticipation." When they checked the signal, however, it turned out to be nothing but a short circuit.

Having tried everything imaginable on earth, Alvarez looked to the moon. In June 1969, the Apollo astronauts would return with moon rocks, which had been continuously exposed to cosmic rays for billions of years. Alvarez obtained permission to test them, as long as they weren't ground up or spun at high speeds. The samples would be rotated repeatedly on a belt through a superconducting coil, with the current increasing slightly with every pass if a monopole was there.

Working three shifts a day at the Lunar Receiving Laboratory in Houston, Alvarez's team tested eight kilograms of moon dust, or about a third of the total brought back to earth, loading the samples in a glove box in a vacuum to avoid exposure to unknown pathogens. Once again, they found nothing, leaving Alvarez doubtful that a monopole would ever be detected in nature.

Although the experiment had fallen short, it testified to his connections at NASA. He hoped to continue his search for antimatter in space, using a satellite from the High Energy Astronomy Observatory program, but even after he scaled it down, the apparatus was still large and heavy. In the end, the agency prioritized the Viking missions to Mars, and his plans were never put into practice.

He had reason to feel concerned about government support of his work. In early 1973, the President's Scientific Advisory Committee was eliminated. Among the dismissed advisors was Alvarez, who had supported Richard Nixon in the last election, "after voting against him three times." He was equally discouraged by the "shabby, disgusting, immoral performance" of Watergate. As Nixon's crimes became clear, he said, "I'm glad at the moment that he *did* fire us, because it would be very embarrassing at the present time to be part of the administration."

Alvarez had voted for Nixon largely out of admiration for the reopening of relations with China. Given his family history—his mother had been born in Fuzhou—he had lobbied to travel there himself, and he was finally invited to join a delegation of physicists. In June, Alvarez and Janet arrived for a highly choreographed visit of three weeks. The trip was notable mostly for his efforts to introduce the Chinese to the frisbee, which he demonstrated in Tiananmen Square.

On July 11, 1973, Alvarez received a telegram in Beijing notifying him that his mother had died. In his diary, he wrote, "One can't be too sad when a loved one dies within a few months of ninety years of age, in full possession of her mental faculties. My mother's mother had a stroke that left her speechless for five long hard years during which time she lay on a bed and stared straight ahead, without indicating much more than her sense of great shock that the Lord would have done such a terrible thing to her, after her long years of service to Him, in the missions of China."

AFTER RETURNING HOME, Alvarez prepared to move into a new phase. At sixty-two, he still earned high ratings as a teacher, but he no longer enjoyed it as much as before, so he decided to retire from the classroom. He would keep training graduate students and hosting his Monday night seminars, hoping to spend the period before his full retirement in the equivalent of a sabbatical "in residence." With his salary paid by NASA, he looked forward to concentrating on astrophysics: "I plan to learn many new things in the next four years."

Alvarez continued to show up every day at the lab, where McMillian had retired as director. In his office, a translucent celestial sphere shared space with a model of a pyramid, while the shelf by his desk was

Alvarez at his desk in 1968. Framed photos include portraits of Albert Einstein, Don Gow, Enrico Fermi, Arthur Compton, Ernest O. Lawrence, Alfred Loomis, and Rowan Gaither. Not visible are pictures of Ernest Rutherford and Walter C. Alvarez.
Lawrence Berkeley National Laboratory.

lined with photos of his heroes, including his father and Don Gow. One portrait reflected his own concerns about aging: "I have an intense dislike of pictures of the white-haired Albert Einstein. . . . I have a picture of the *real* Einstein on my desk—a very young patent examiner who did marvelous things in his spare time."

Many of his colleagues later recalled Alvarez's boyish side: "We remember our occasional delight when opening Luie's office door to find him doing a handstand on his desk." He enjoyed parlor tricks, like keeping a dime balanced on the hook of a spinning coat hanger, and was an early adopter of the elastic Super Ball, showing how it returned to his hand when he tossed it between a table and the floor. His house was full of puzzles, and he was disappointed when students lacked the dexterity to solve them as quickly as he could.

Unlike most professors, he always ate lunch with younger researchers, whom he saw as soldiers on the front lines. To look for potential protégés, his older colleagues served as talent scouts, bringing over prospects for Alvarez to evaluate in the cafeteria. By the seventies, his weekly seminars were dominated by students with long hair and beards. Although his politics were far removed from the mood in Berkeley, he showed grudging respect for the Free Speech Movement: "Their goals are without merit, but their tactics are brilliant."

As always, he was extremely regular in his habits. He bought half a dozen identical sports coats at a time, and in a laboratory setting—where he always reminded visitors to put their hands in their pockets—he rarely departed from his usual outfit of brown shoes, blue slacks, and a shirt with short sleeves. His breakfast was invariably an egg and English muffin, and in the cafeteria, Jack Lloyd, who oversaw logistics for the balloon program, never saw him order anything but "a hamburger with a lot of ketchup and a Heineken."

In books, he preferred nonfiction, telling the psychologist Anne Roe, "I used to read *Astounding Stories* but it's not one of my vices." A fan of classical music, he claimed that he could pick out anything on the piano in the key of F. Although he sometimes played pieces for four hands with Janet, his children mostly heard endless renditions of Scott Joplin's "The Entertainer."

Every night, he watched the evening news for problems to solve, scribbling ideas on napkins. Colleagues learned to expect calls from him at dinnertime, asking why his latest brainstorm wouldn't work. His election to the National Inventors Hall of Fame, he later said, meant as much as the Nobel Prize, and invention itself was a habit that he diligently maintained: "If you can't think up something really classy, you think up something trivial."

Alvarez was darkly amused by a lawsuit filed against the Nobel laureates Emilio Segrè and Owen Chamberlain by the physicist Oreste Piccioni, who claimed that they had failed to credit him for the idea that led to the discovery of the antiproton. Although it was dismissed on a technicality, Alvarez believed Piccioni: "I had long wondered how Emilio had devised the beautiful experimental setup . . . as it was so far from anything I had seen him do."

He was often reminded of his past. After Lewis Strauss died in 1974,

Alvarez wrote to his widow: "I considered Lewis to be a close personal friend, as well as a good friend of science. . . . His personal support for my research projects was a key element in their success." Recalling their summers at Bohemian Grove, he concluded, "Lewis was one of the finest men I ever met, and I will miss him."

Another memorable dispute began in the summer of 1975, when he entered his office to find a colleague rifling through his monopole file, "to which my secretary had quite properly given him access." At forty-two, Buford Price resembled a younger version of Alvarez himself, with a gift for ingenious experiments in a wide range of fields, including astrophysics.

Price's most notable innovation was the "track etch" technique. When a charged particle passed through a stack of plastic sheets, it left an invisible ionization trail. Using a chemical solution, like lye, the damaged material was dissolved to reveal a microscopic track. It was more practical for balloons than Alvarez's unwieldy magnetic spectrometer, and Price used it to look for heavy nuclei in cosmic rays, flying a detector for sixty hours above Sioux City, Iowa.

After two years of analysis, Price focused on one anomalous event. Something had penetrated over thirty sheets of plastic, making a series of conical holes. Most particles would have halted before reaching the stack's base, indicating that it was truly massive—around two hundred times the mass of a proton. Purely by accident, Price concluded, he had found a magnetic monopole.

As word spread of what appeared to be "one of the major scientific events of the century," Price announced it at a press conference in August, supposedly to get ahead of the rumors. When he finished answering questions, Alvarez took his place at the podium. On hearing the news, he had shaken Price's hand, but he also raised doubts in informal discussions. Now he publicly voiced his suspicions that the track had been made by something else entirely.

Faced with a dispute that might tear the lab apart, Alvarez looked carefully at Price's data. From his own unsuccessful searches in materials that had been bombarded for millions of years, he was skeptical that a monopole would have struck the detector completely by chance. An erratic motion in the lower layers also seemed inconsistent with a monopole, which wouldn't deviate from its original path. Price, he thought, had rushed to announce it without asking the hard questions first.

As Alvarez's children later recalled, he acted as though he were "going into battle." After crunching the numbers for weeks, he determined that the event could have been produced by a heavy platinum nucleus, which fragmented into osmium and finally tantalum. Price had ruled this out because of the stack's thickness—a platinum track wouldn't have made it to the bottom.

Playing a hunch, Alvarez called Ed Hungerford, Price's collaborator at the University of Houston, who informed him that the stack was only two-thirds as thick as initially reported. In his analysis, Price had assumed, incorrectly, that the detector was identical to one from an earlier flight. When the true dimensions were used, Alvarez's alternate scenario—which undermined the entire claim—was perfectly plausible. In a call with Dirac, whose work had inspired the monopole hunt, Alvarez declared himself "completely dissatisfied" with Price's conclusions.

With a smoking gun in hand, he weighed his options. Remembering the disappointment of his single "antioxygen" event, he knew how hard it could be to walk away from a major discovery. He called Price, who was at a conference in Munich, and spent an hour arguing that they should jointly publish the platinum hypothesis. Price declined, stating only that he would check his calculations and send a telegram if he turned out to have been wrong.

Feeling that Price had failed an important test, Alvarez decided to make an example of him. A physics conference was underway that week at the Stanford Linear Accelerator Center. Although the deadline for proposals had already passed, Alvarez phoned the organizers, saying that he wanted to deliver a presentation at the final session without announcing a topic in advance. In a clever piece of stagecraft, he finally convinced them to schedule the talk "blind."

On August 27, 1975, Alvarez sat expectantly with Janet in the front row of the auditorium. Word of his plans had spread, and when he rose to speak, the room was packed. In devastating detail, he demonstrated that Price had failed to rule out alternatives "as vigorously as possible." He closed by thanking his absent opponent: "What otherwise might have made for a tense situation—no one really likes to have his firmly held conclusions questioned—was ameliorated by the fact that Buford and I are friends and longtime respected colleagues."

When he was finished, the session chairman read a cable from Price, confirming that Alvarez was right—the stack was indeed thinner than had been previously thought. Basking in his victory, Alvarez headed for the cafeteria, where he saw Richard Feynman seated at a table. As he entered, Feynman rose to meet him, exclaiming, "That was a great detective job!"

Alvarez was gratified, but the dispute was far from over. At first, Price refused to retract, saying only that he would review the data, and even after he conceded, he felt that it was "strictly not cricket" for Alvarez to have acted in such an ostentatious fashion: "He saw a circus in the making. He was responsible in part for the huge crowd and the circus atmosphere."

Believing that he had given Price a fair chance, Alvarez said afterward that their friendship was "a bit strained," which was a considerable understatement. Privately, however, he made his feelings clear. Years later, during an even more heated debate, he took an opponent aside to offer a word of caution: "Let me warn you. Buford Price tried to oppose me, and when I finished with him, the scientific community pays no more attention to Buford Price."

WHEN ALVAREZ SAID THAT his balloon program had attracted "young graduate students that I think are the best available," he almost certainly had a specific person in mind. Richard A. Muller was raised in the South Bronx and arrived in Berkeley in 1964. He initially met Alvarez as a head teaching assistant, and on his first visit to the lab, which entranced him, he made a good impression. Without even looking at his academic record, Alvarez asked, "When can you start?"

Muller was thrown into the deep end. Alvarez advised him to forget about any background reading and head straight to Building 46. "Just go over there and hang around. Do anything anybody asks you to do. Sooner or later someone will see that you're there, and they'll ask you to hold a screwdriver. Get your hands dirty. Pretty soon you'll know how things are constructed."

A few weeks later, Muller was helping out at the lab when he dropped a photomultiplier tube, which exploded as it struck the floor. On learning that he had destroyed $15,000 of equipment, Alvarez only stuck out his hand, congratulating Muller on breaking his first expensive piece of apparatus. "Now I know you're becoming an experimental physicist."

Alvarez came to see Muller as his scientific son. The man he praised as "certainly the best student I have had" became the first person he called to ask how his latest idea might be wrong. Muller, in turn, declared that he wanted to learn how Alvarez had achieved so much, which was exactly the right thing to say. Even in his own group, Alvarez's wilder notions were often dismissed at first, just as his associates in radar had learned to wait until he made the same suggestion three times. Muller took every idea seriously: "I assume that everything he said was thoughtful and deep."

When Alvarez asked why Muller had declined an assistant professorship at Harvard, he explained, "I'm not finished learning from you." As Muller recalled, "[Alvarez] had learned just enough about every subject; he could go back and fill in the gaps later, when and if that was necessary. The gaps in his knowledge were surprisingly large, but not detrimental to his work." This approach made it possible to see new connections, but only if you looked beyond the obvious: "What I learned from Alvarez is that the best projects aren't the ones that have drawn dozens of people—they're in the valleys between the peaks."

Alvarez did this methodically, combining articles from old journals—"he referred to them constantly"—with what he had learned that week. Every Friday, he still followed his father's advice "to think crazy" in the afternoon. "Only one out of ten ideas, he said, were worth pursuing," Muller remembered. "Only one of ten of these would last a month.

Only one out of ten of these would lead to a discovery. If these figures are true, then Luis must have had tens of thousands of ideas."

Muller never forgot a conversation that they had in the seventies, after thousands of faulty welds were discovered in an Alaskan oil pipeline. After inquiring if Muller had heard the news, Alvarez asked, "Have you solved it?" He then described his own idea for guiding repairs with specialized X-rays. It was never used, but Muller was astonished: "I had the realization that I had the capability as a physicist to invent things that might help society."

Alvarez also taught him that while both scientists and ordinary people could fool themselves, a scientist took steps to prevent it, knowing that "nothing looks so much like a new effect as bad technique." The real scientific method wasn't hypothesis testing, but a determination to catch potential errors. As Alvarez said in his autobiography, "Most people are concerned that someone might cheat them. The scientist is even more concerned that he might cheat himself."

For anyone who fell short of that high standard, Muller saw how brutal—even terrifying—he could be: "He would pull down your pants in public." Alvarez expected others to admit their mistakes at once, and he didn't hesitate to yell within earshot of the entire lab. Muller rarely received this treatment, but he understood the lingering pain that it caused, while emphasizing the rewards for those who endured it: "People tolerated Luie Alvarez because it was so exciting to do physics with him. I would call him the greatest scientist of the last hundred years."

By 1972, Muller felt ready to propose his own project. His colleague George Smoot recalled Alvarez's advice: "Before you rush into continuing with a new, improved version of the experiment you've just done, sit down and take stock. . . . Think particularly deeply about the important opportunities in physics, opportunities that have emerged thanks to new technologies and recent experimental results. Take a month or two to think about it; study what the most interesting and important problems are. Then decide what you want to do next."

Looking for ideas, Muller remembered an experiment, proposed by the astrophysicist Jim Peebles, involving the cosmic microwave background. All of known space was filled with a faint radiation that was evidently a remnant of the period just after the Big Bang. Because of its common origin, it seemed highly isotropic, or uniform, but basic physics implied that it should really be slightly anisotropic, which meant that its properties would vary based on where it was measured.

As the earth passed through the cosmic microwave background, Peebles noted, the waves in front were compressed, so their frequency should become somewhat higher, much as the Doppler effect caused a passing siren to rise and fall in pitch. "Just as more water hits your face than the back of your head when you run through rain," Muller wrote, "the intensity of the microwaves should be slightly greater in the forward direction."

Muller wanted to look for anisotropy in the microwave background, which might also emerge from primordial variations in the distribution of matter in the universe. Alvarez doubted that they would find the latter, dismissing the effort as "a waste of time." After Muller pointed out that they would at least see the effects of the earth's motion, Alvarez brightened: "Wow, that's really interesting. Okay, you *will* have something. Let me help you raise money for this."

After recruiting George Smoot, Muller set to work. The observations would be affected by microwave radiation in the atmosphere, which they needed to subtract from their measurements. This would be easier at higher altitudes, but neither of them had any desire to mess around with balloons again. The U-2 spy aircraft was more than capable of reaching seventy thousand feet, so Alvarez drew on his networks to secure support from Lockheed and NASA.

The flights began on July 7, 1976. Carrying a special radio receiver, a U-2 flew back and forth in parallel paths, building a celestial map that disclosed the anisotropy that they had expected. There was just one problem—it was in the wrong direction. The only explanation, they concluded, was that the Milky Way itself was moving at over a million miles per hour, drawn by gravitational attraction from a gigantic unseen region in intergalactic space.

If there was a concentration of mass—later nicknamed the Great

Attractor—massive enough to affect the entire galaxy, there had to be others. Determining their origins would require even more precise measurements of the microwave background, in search of the fluctuations that had allowed large structures to appear after the Big Bang. To completely eliminate atmospheric interference, the data had to be taken by a satellite, which became known as the Cosmic Background Explorer.

Long before that point, however, Muller had left the project. He told Alvarez that after their anisotropy paper was published, his relationship with Smoot had rapidly degenerated. When NASA assembled a team to study the proposal, Smoot—who was willing to remain with the program for the years that it would take to realize its goals—was joined by John C. Mather, an astrophysicist at Goddard Space Flight Center, but Muller was no longer involved.

Mather was privately relieved that Alvarez had also declined to take a role. "Luie was considered brilliant, creative, and terrifying. He was known for unmercifully battering his postdocs and other scientists during seminars and oral presentations. I attended a meeting in Berkeley once when Luie tore into another scientist, a physicist named Mike Lampton. . . . I decided right away that Luie was not a person with whom I would enjoy working."

Muller was less interested in navigating the bureaucracy of NASA—he described his withdrawal as "one of the great decisions of his life"—than in exploring a different problem with Alvarez. Years earlier, Murray Gell-Mann and George Zweig had independently proposed the quark, an elementary constituent of matter that formed composite particles, including protons and neutrons. No one had ever seen it in isolation, however, supposedly because it was confined to the nucleus.

Alvarez suspected that everyone was looking in the wrong place. Most experimenters assumed that quarks had either –1/3 or 2/3 of a proton's electrical charge, but if they had a charge of exactly one unit, they could easily be confused with different particles. He proposed searching for "integral" quarks by using the cyclotron as a mass spectrometer, just as he had done when he accidentally found helium-3 in 1939. Filling the chamber with hydrogen, they could tune the magnetic field to accelerate various masses, looking for a very small particle with the charge of a proton.

"Luie thought that we could complete the experiment in a few months," Muller wrote. "By 'we,' I knew Luie meant me. He had recognized the importance of the measurement and figured out how to do it, and it would be my job to do the detailed experiment design and to make the measurements." It was a test of loyalty, as well as a chance to master the cyclotron, but only if he scaled back elsewhere. "I guessed as best I could the probability of a major discovery in the quark search and somehow came up with the figure of 10%. I weighed this probability against the importance of such a discovery and decided to take the risk."

Two years later, they had found nothing. The quark of unit charge didn't exist. "I had taken a calculated risk and had lost," Muller recalled, although it turned out to be far from a wasted effort. One day, he learned that another physicist was using a similar approach to look for superheavy elements. He felt chagrined that he had failed to think of it, despite spending countless hours at the cyclotron: "I had been lazy and had become too narrowly focused on the details of one experiment."

Looking for other applications, Muller remembered something else that Alvarez had brought to his attention. Alvarez's old friend Willard Libby—now married to Leona Marshall—had won a Nobel Prize for the invention of radiocarbon dating. Living plants and animals absorbed trace amounts of carbon-14, a radioactive isotope produced by cosmic rays, which decayed at a constant rate after the organism's death. By seeing how the relative abundance of radiocarbon had fallen, archaeologists could date organic materials with unparalleled precision.

At first, carbon-14 was measured by detecting decay events, which were hard to see in very small or old samples. It occurred to Muller that by using the cyclotron as a mass spectrometer, he could simply count the atoms directly. When Alvarez walked into his office, Muller told him that he knew how to increase the sensitivity of carbon dating. "As was customary between Luie and myself, I didn't tell him my solution immediately; I gave him a moment to try to invent the method himself. Luie paused and I worried that he would reinvent my idea on the spot."

After finally asking Muller for the answer, Alvarez extended a hand and said, "Congratulations!" Alvarez later expressed admiration for the insight: "I was aware of all the necessary ingredients, including the importance of the problem, yet I failed to bring them together." Before long, it became the dominant form of radiocarbon dating. With a sense of audacity that evoked Alvarez at his best, Muller planned to date the Shroud of Turin, the alleged burial cloth of Jesus. He never got the chance, but investigators used his technique, over a decade later, to determine that the shroud was a forgery from no earlier than the thirteenth century.

Muller's confidence had been reinvigorated. For his next major project, he turned to another problem that would meet Alvarez's high standards—the question of whether the universe would expand forever or eventually contract in a cataclysmic Big Crunch. To see if its expansion rate was slowing, he needed to compare the velocity of near and distant galaxies, which called for a reliable method of measuring how far away a particular galaxy might be.

In 1978, Muller heard a talk by the cosmologist Robert Wagoner, who described using supernovas as "standard candles." By comparing the known luminosity of an exploding star with its apparent brightness from earth, it was possible to calculate its distance. The hard part was obtaining a large enough data set. Supernovas were usually found by taking photos of a region of space at different times, searching for new points of light, and the process was agonizingly slow.

On learning about the problem, Alvarez was reminded of a project initiated in the sixties by the astronomer Stirling Colgate, who built a telescope that automatically scanned for supernovas. Colgate had hoped to use computers to analyze the images, but the technology of the time was too primitive. Alvarez reasoned that advances in computing and cameras would make it more practical now, just as similar improvements had enabled his breakthroughs at the bubble chamber.

To help run the project, Muller hired a young astrophysicist named Carl Pennypacker, who learned to appreciate one of Alvarez's favorite sayings: "If you ever knew how hard an experiment would be before you went into it, you would never do anything." Drawing on his connections and experience, Alvarez advised them not to wait until the computers were ready, but to start at once with human scanners. Pennypacker recalled his insistence on chasing big goals, even at risk of failure: "If your junk pile is not big enough, you are not trying hard enough."

All the while, Alvarez was winding down his own obligations. At the beginning of 1978, he took early retirement, slightly ahead of schedule, "to escape from my very restrictive patent contract." Six months later, Walter C. Alvarez died at ninety-three. He had been an inescapable presence for his son, who measured himself against his father's accomplishments. His other mentor, Alfred Loomis, was also gone, and with no comparable figure remaining from his formative years, it might have seemed like a good time to ease into a peaceful final chapter.

On September 22, 1979, one of three satellites from Project Vela—an American surveillance system that monitored nuclear weapons tests—registered an anomalous reading. The satellite carried a pair of sensors, known as bhangmeters, designed to detect the characteristic double flash of an atomic explosion. At the instant of detonation, the air around a bomb became incandescent. The ensuing shock wave compressed the atmosphere, briefly making it opaque enough to block the fireball, which was finally revealed in a second surge of brightness.

Two peaks were visible on the satellite readout. An air force analysis pointed to a detonation at around three in the morning, somewhere in an area encompassing the Indian Ocean and the African coast. In his diary, President Carter wrote, "There was indication of a nuclear explosion in the region of South Africa—either South Africa, Israel using a ship at sea, or nothing."

Since the Carter administration was publicly committed to arms control, any nuclear test threatened to undermine the delicate international situation. After word of the alleged detonation was leaked, the geophysicist Frank Press, the president's science advisor, convened an investigative committee of nine scientists, including Alvarez, Muller, and Pief Panofsky, with the MIT professor Jack Ruina—a longtime member of Jason—serving as chairman. According to Alvarez, its real driving force was Hans Mark, the secretary of the air force, who wanted "a second opinion" about an event that he was wary of attributing to a bomb.

The panel would evaluate whether the signal was a false alarm, a natural phenomenon, or a nuclear detonation. At first, most of the members were inclined toward the third explanation, but the data seemed inconclusive. The two bhangmeters had registered different intensities for the second peak. While this could have been caused by satellite movement or a technical malfunction, it might also indicate that the source of the flash was physically closer to one sensor than the other, which would mean that it was near the satellite itself, not thousands of miles away.

Drawing an analogy to his work with the bubble chamber, Alvarez asked the Defense Intelligence Agency for a collection of "zoo-ons," or previous readings "so strange they belonged in a zoo." Along with Muller and the physicist Richard Garwin, he carefully examined hundreds of records from computer tapes. A few earlier events, they found, had some of the characteristics of a detonation, but could be ruled out for other reasons. Given a large enough range of anomalous readings, Alvarez thought, eventually one would look like a bomb.

To his eyes, the Vela incident didn't resemble the twelve authenticated tests detected by the satellite, and corroborating evidence had yet to emerge. Despite repeated flyovers, no radioactive debris was detected in the atmosphere. While its absence might have been due to rain at the test site, this wasn't consistent with the bright flash, unless a break had opened in the clouds at just the right time.

He dismissed other alleged proofs as "a wild assemblage." High levels of radioactive iodine had been found in the thyroids of Australian sheep, a data point that he described as "statistically insignificant, but which had excited the men who made the measurements." The Arecibo telescope in Puerto Rico recorded an ionospheric disturbance that was consistent with a test near South Africa, but he didn't think that they had enough of a baseline to draw any conclusions.

Perhaps the most persuasive clue came from the Sound Surveillance System, the classified submarine detection array implemented on the recommendation of Project Hartwell—the advisory committee on which Alvarez himself had served. On the day of the Vela incident, it picked up acoustic signals that the panel saw as ambiguous, although a later study determined that they pointed to a test near the Prince Edward Islands, a South African territory in the Indian Ocean.

In the end, the committee favored the hypothesis that the satellite signal was caused by a collision with a micrometeorite. Some of the reflective debris could have passed into the field of view of the bhangmeters, producing the first pulse, followed shortly afterward by a second wave of particles. Describing this alternative as "reasonable" but not "necessarily correct," their final report, which was released in 1980, concluded that the reading "was probably not from a nuclear explosion."

Alvarez put it even more strongly in his memoirs. After stating, incorrectly, that "only one of the two satellite sensors" registered the event, he said, "I doubt that any responsible person now believes that a nuclear explosion occurred because no one has broken security." In fact, at least one "responsible person"—President Carter himself—continued to believe that the Vela incident was a nuclear test, and there were legitimate reasons for principled observers to reach the same conclusion.

Military intelligence pointed unequivocally to nuclear weapons development by Israel and South Africa, and a defense report found that the Vela signal was "very unusual" in comparison to other members of the Alvarez "zoo." The "phase anomaly" between the sensors could plausibly be explained as a malfunction on the satellite, which was beyond its operational lifetime. Houston T. Hawkins, a security scientist at Los Alamos, stated that similar issues "had been seen on all of the more recent nuclear explosions that had occurred in [the satellite's] field of view."

Dieter Gerhardt, a South African naval commander who was later exposed as a Soviet spy, claimed to a reporter that the incident had indeed been a joint test with the Israelis, code-named Operation Phoenix. Another source told the historian Richard Rhodes that the site was furnished by South Africa, while the journalist Seymour Hersh quoted an Israeli official who blamed a "fuck-up" for their detection: "There was a storm and we figured it would block Vela, but there was a gap in the weather—a window—and Vela got blinded by the flash."

Critics of the Ruina Panel openly wondered if it had reached its verdict to avoid embarrassing Carter. At times, Ruina himself seemed to encourage this viewpoint: "My job was to see if there was any other explanation than a nuclear explosion." Panofsky, by contrast, described the panel as "uninstructable," while Muller, who forcefully defended its conclusions, said bluntly, "People who know these scientists find the concept of political pressure laughable."

Alvarez was unlikely to have agreed to slant his findings, but he had reason enough to tread carefully when the stakes were so high. His talent for finding plausible alternatives had served him well, but in the aftermath, as usual, he came down hard on one side, as if only a fool would think that any doubt existed. Rhodes felt that the search for an alternate scenario "overcame his usual good sense," given the nature of the assignment: "If Alvarez had personally investigated the Vela event, my guess is he would have concluded that it registered a nuclear test."

IN 1967, ALVAREZ HAD EXCHANGED LETTERS with a Kennedy conspiracy buff who eventually tested his patience. Moving to end the discussion, he wrote, "I have no further interest in the Kennedy assassination films." Two years later, however, he received a call from a lawyer for Clay Shaw, the New Orleans businessman arrested by District Attorney Jim Garrison on charges—of which he was later acquitted—of conspiring in the president's murder.

When Shaw's legal team asked for copies of his correspondence with CBS about the shots in Dealey Plaza, Alvarez declined. Although he continued to believe that only three bullets had been fired, he conceded that a knowledgeable opponent could still find flaws in his argument. In all likelihood, he was thinking of Josiah Thompson, a Yale graduate with a background in philosophy, who had conducted research for the *Life* article that inspired Alvarez's work.

On January 1, 1969, Thompson—who outlined his theories in the book *Six Seconds in Dallas*—had sent along a critique of Alvarez's analysis of the blurs in the Zapruder film, noting that they could be taken as evidence for four shots, not three. He also emphasized that the suggestion of a siren before the final shot was contradicted by Samuel Kinney's testimony.

In his response, Alvarez ignored most of Thompson's points, stating that he was glad that he had helped "to persuade the public that the assassination critics are a bunch of nuts." To another correspondent, he wrote, "I once leafed through a copy of Thompson's book, when I was in the San Francisco airport bookshop. . . . In his mind I am simply an idiot who could not see some blurs on some photographs that he noticed after I had called attention to the phenomenon."

All the same, Alvarez was bothered by the recognition that his argument was unlikely to hold up in court. A week after speaking to Shaw's attorneys, on February 27, 1969, he left for the American Physical Society meeting in St. Louis, bringing a copy of *Six Seconds in Dallas*. His graduate student Paul Hoch had advised him that its analysis of the backward snap of Kennedy's head was a compelling argument for a shot from the front—and therefore for a conspiracy.

During a night "when I couldn't get to sleep," Alvarez came up with an alternative explanation that used Thompson's charts against him. Scribbling equations on a cocktail party invitation, he worked out the jet effect hypothesis—the idea that the momentum of the matter blown out the front of the skull exceeded that of the bullet's impact at the rear, driving Kennedy's head back and to the left.

On June 29, 1969, his associate Sharon "Buck" Buckingham went to the San Leandro Rifle Range to shoot targets supplied by Alvarez, including five melons, three coconuts, and two water jugs. The range master, who was eager to leave for the day, was skeptical: "I've been around guns all my life, and you must be out of your mind to believe something you hit with a bullet will come back to you."

Buckingham's initial tests did nothing to silence the range master's doubts. Decent results were obtained from smaller melons covered in filament tape, which moved a few inches toward the shooter, but the other targets failed to support the hypothesis. Coconuts and a jug of gelatin were driven in the same direction as the gunshot, while a container of water ended up a full six feet downrange.

On the afternoon of February 15, 1970, Alvarez participated in a second trip to San Leandro, but it was equally discouraging. Rubber balls of gelatin tended to move away at an angle; plastic water bottles simply burst; and a pineapple sent its largest piece flying perpendicular to the lane. None of the targets reacted as he had anticipated. In May, however, he conducted a successful demonstration that exclusively used small melons, which were the only objects to exhibit retrograde motion.

Thompson later argued that there were serious issues in applying these results to the assassination. "Whether a melon is taped or not, a bullet will cut through its outside like butter. A human skull is completely different. Penetrating the thick skull bone requires considerable force, and that force is deposited in the skull as momentum. The Alvarez theory requires little momentum transfer at entry and a great blowout at exit." In government tests at Aberdeen Proving Ground, skulls filled with gelatin had all moved in the direction of the bullet.

Another problem was the ammunition. Alvarez's marksmen used soft-nosed hunting rounds that mushroomed on impact, but the bullets in the assassination were jacketed in copper, allowing for greater penetration—a crucial element in the "magic bullet" theory. As Alvarez admitted, this fact "was apparently important in intensifying the explosive jet effect." Since Buckingham had also loaded his cartridges by hand to increase their velocity, Thompson noted, each one "would have struck its target with about three times the force of a bullet from Oswald's rifle."

Alvarez, revealingly, waited for five years to write up the results. His daughter Helen, who was seven at the time, vividly recalled wandering one Saturday into the living room, where her father was watching the fatal headshot on a loop. In 1976, he published what he called "a documented counterexample . . . to disprove the assertions of many [assassination] writers concerning the consequences of Newton's laws of motion." And by describing only the final melon test, he made it look as though nearly all the targets—not fewer than half—had reacted as predicted.

What both Alvarez and his subsequent critics tended to ignore was that the backward movement of Kennedy's head was greater than could be produced by any gunshot, even from the front. As Larry Sturdivan, a physical scientist from Aberdeen Proving Ground, explained:

> The deposit of momentum from the bullet is not sufficient to cause any dramatic movement in any direction. . . . It would have a slight movement toward the front, which would very rapidly be damped by the connection of the neck with the body.
>
> In other words, the head would begin to move and then the body would be dragged forward with it at a much lower velocity. Certainly not a very large velocity. Not throwing anybody anywhere.
>
> In fact, I conclude from these films that, since the president does have motion, that it must have arisen from another source, that is, it could not have been the momentum of the bullet.

The inevitable conclusion was that the head movement had no value in establishing the shooter's location. Many experts believed that it was actually due to a delayed neuromuscular reaction, produced by a traumatic brain injury, which did nothing to help the case for either direction. Physics had no more advantage here than it did with the Zapruder film.

But Alvarez's most extensive engagement with the assassination was still yet to come. It emerged from the House Select Committee on Assassinations, which was formed in 1976 to investigate the killings of Kennedy and Martin Luther King Jr. After two years, it was preparing to confirm that Oswald acted alone, only to be disrupted at the last minute by a bombshell development.

Through the conspiracy community, the HSCA acquired recordings of Dallas police radio from November 22, 1963. It included about five minutes of uninterrupted audio, evidently produced when a motorcycle officer pressed a button to talk, leaving it stuck on transmission mode. No one knew exactly where or when it was taken, but it was possible that it had picked up the gunshots.

Since it was hard to hear anything but an engine, the committee sent it to Bolt Beranek and Newman, a research firm in Cambridge, Massachusetts. To obtain test shots for comparison, microphones were placed along the motorcade route, recording rifles from the Book Depository and the grassy knoll. The chief scientist, James E. Barger, looked for corresponding patterns in the police audio, eventually identifying four "impulse sounds." Three seemed to come from the Book Depository, while the odds of a fourth shot from the grassy knoll were "about even."

Given the enormous importance of the answer, the committee asked two audio experts, Mark Weiss and Ernest Aschkenasy of Queens College, to check the results. They reviewed only the supposed grassy knoll shot, which Barger thought was the third one fired. Using a map, they calculated how long it would take for the sound of a shot—and its echoes—to travel to the position of a radio in the motorcade. Finding one impulse sequence that was "very similar" to this prediction, they put the probability of a gunshot at over 95 percent.

What remained unresolved was whether a motorcycle had even been in the right place to capture the sounds. The committee thought that the most likely candidate was Officer H. B. McLain, who doubted that he was responsible, while photos seemed to place him farther back in the motorcade. Another issue arose from the two different radio channels used by Dallas police. The alleged shots were captured on Channel 1, but McLain insisted that he had been on Channel 2, like every officer on the parade route. Finally, the recording lacked any trace of crowd noise or the sirens that followed the fatal shot, which should have been audible in Dealey Plaza.

Despite these objections, the apparent credibility of the acoustic analysis was powerful enough for the committee to throw away years of work. In its initial report, released on December 30, 1978, the HSCA stated that "scientific acoustical evidence established a high probability that two gunmen fired at President John F. Kennedy," indicating that the president was "probably assassinated as a result of a conspiracy." It was the same desire for objective proof that drew Alvarez to the Zapruder film, but with diametrically opposite results.

In August 1980, the Justice Department asked the National Academy of Sciences to advise on whether the case should be reopened. Philip Handler, the academy president, wanted Alvarez to chair the investigation. "Since the buffs would automatically have rejected any report published under my name," Alvarez recalled, "I agreed to be a committee member but suggested Norman Ramsey"—a Harvard professor who had worked with him at the MIT Rad Lab and on the limited-warfare study—"as a competent and acceptable chairman."

On October 20, Alvarez finished a critique for the Committee on Ballistic Acoustics. Calling the Weiss and Aschkenasy report "particularly shoddy," he noted that they hadn't checked any of the confirmed shots from the Book Depository, which would have tested the method's reliability. It was easy, he said, to find correlations "between two quite unrelated sources of noisy sound."

The Bolt Beranek and Newman analysis struck him as slightly more professional, but this was only "damning it with faint praise," and one detail was especially damaging. Its first two impulses—both supposedly gunshots by Oswald—were only 1.66 seconds apart. Tests with the assassin's rifle had established that it was all but impossible for one person to fire that quickly.

And an even more devastating takedown would soon follow. Its unlikely source was Steve Barber, a twenty-five-year-old conspiracy buff and occasional rock percussionist from Ohio. The previous summer, he had bought an issue of the adult magazine *Gallery* that came with a playable record with audio from Channel 1, advertised as the "evidence that destroyed the lone assassin theory."

On September 12, 1980, Barber was listening to the recording when he noticed something for the first time. Shortly after the first alleged shots, he heard two very faint words: "Hold everything." Wondering what it meant, he suddenly made a connection. A friend had sent him a bootleg cassette of Channel 2, which he inserted into his tape player, listening for a familiar phrase. At one point, Bill Decker, the county sheriff, had radioed the dispatcher: "Hold everything secure until the homicide and other investigators can get there."

Decker's transmission was made a minute after the assassination, so it couldn't have been recorded at the same time as the gunshots. Barber eventually concluded that it was an example of "crosstalk." When two radios were in close proximity, it was possible for one to pick up sounds from the other, like any form of background noise. In this case, a radio had been broadcasting loudly enough on Channel 2 to be heard on a second radio nearby, which was tuned to Channel 1.

This discovery was enough to destroy the HSCA's conclusions. There was nothing in the Channel 1 recording by itself to establish the exact time of the impulses, which were identified as shots after the fact. The simultaneous crosstalk from Channel 2 implied that it had really been recorded about a minute later. If that was true, then the sounds couldn't be gunshots at all.

Barber promptly informed the Ramsey Panel that he had found evidence that could save them "much time and money." In November, a staffer forwarded his letters to Alvarez: "I think [Barber] sounds his horn too loudly but on the other hand I should not make that judgment." Seeing the significance immediately, Alvarez told Ramsey, who supported both the "Hold everything" identification and another possible piece of crosstalk that sounded like "Chaney."

On January 31, 1981, the panel questioned the audio experts. Alvarez felt that James Barger was frustratingly evasive: "I realized then that a professional expert witness can't afford to admit that he has made a mistake." The members soon concluded that the acoustics studies—which arrived at different times for the grassy knoll shot—were both deeply flawed. According to Barger, Alvarez didn't mince words: "He didn't care what I said, he would vote against me anyway."

When they turned to the crosstalk angle, they saw that the "Hold everything" and "Chaney" identifications couldn't both be right. James Bowles, the communications supervisor of the Dallas police, who prepared careful transcripts of the recordings, believed that "Chaney" was actually the phrase "I'll check it." He later matched this with a statement on Channel 2 by Deputy Chief N. T. Fisher, which coincided exactly with the assassination.

The panel faced a stark choice. "I'll check it" seemed to support the timing of the disputed acoustics analysis, while "Hold everything" would disprove it entirely. Working with the FBI, they prepared spectrograms, or voiceprints, of both alleged crosstalk segments on Channel 1, which could be compared to the Channel 2 phrases to see which one was more convincing.

Alvarez arrived at an objective approach. On a glossy reproduction of the two spectrograms for "Hold everything," he found five pairs of fuzzy bands that looked the same to the naked eye. Marking the center of each band with a sharp tool, he used a millimeter ruler to find the height of the waveform at that point. He calculated that the frequency in one spectrogram was about six percent higher than the other, which matched the known difference in speed between the recording devices for the two channels—a strong indication that the identification was correct.

A more sophisticated computer analysis provided additional confirmation, but the real test was what would happen when the same tools were applied to "I'll check it." If it couldn't be eliminated using identical methods, the analyses would cancel each other out, or at least do nothing to clarify the situation. Fortunately, the results seemed to show that the alleged phrase didn't match the corresponding voiceprint. It wasn't crosstalk, but random background noise.

On November 24, 1981, Bowles wrote to Ramsey, "I'm sorry that Chief Fisher's 'I'll check it' didn't log well." Bowles then revised his transcripts to describe the segment as "discounted as being Deputy

Chief Fisher . . . by sound spectrogram." In the final report, however, the analysis of this snippet wasn't mentioned. The transcript only stated, without context, that "I'll check it" had been "discounted by sound spectrograms," and there was no indication of how closely it had been studied.

The committee evidently recognized that it might still be possible to find reasons to favor the rejected identification, so it deemphasized evidence that was open to multiple interpretations. Alvarez wanted the report—which he felt could be "grossly misused" if it were less than conclusive—to end the debate forever. To put a knife in the heart of the acoustics analysis, he proposed running visual aids on "all three networks," optimistically writing, "My guess is that the matter would then be settled, and the buffs could do something else."

In May 1982, the *Report of the Committee on Ballistic Acoustics* was released. The members concluded unanimously that "reliable acoustic data do not support a conclusion that there was a second gunman," which did little to persuade the conspiracy community. Decades later, Thompson published an entire book hinging on the argument that "I'll check it" was a stronger match than "Hold everything," confirming that the debate over the audio would never end.

Yet common sense alone was enough to cast doubt on the Channel 1 recording. Apart from its other issues, the engine noise implied that the motorcycle was idling or at low speed. Sirens heard much later—two minutes after the "shots"—rose and fell, as if passing a stationary listener. The simplest explanation was that the radio wasn't in the motorcade at all, but three miles away at the Trade Mart, Kennedy's intended destination. Timing himself in his car to estimate the length of the drive, Alvarez found that it was consistent with the delay before the siren sounds.

Ultimately, however, the "Hold everything" spectrogram remained the centerpiece of the Ramsey Panel report, which fell into the same trap as the HSCA. Both treated a single "objective" result as a shortcut to a verdict that actually rested on less tangible factors—basic plausibility, eyewitness testimony, and what was known about Oswald himself.

Alvarez had searched for a similar way out through his Zapruder film analysis and the jet effect theory, neither of which was as conclusive as he implied. Arguments from physical evidence could always be countered by others, while definitive proof was a mirage that remained constantly out of reach.

In any case, Alvarez had washed his hands of the assassination. As far as he was concerned, he had found closure, and he was more interested in another problem that had occupied him for the last two years. His son Walt had brought it to his attention, in the form of a small rock sample with two layers of limestone, sandwiching a thin deposit of clay. Incredibly, it would plunge Alvarez into the final—and most spectacular—adventure of his career, centered on a very different kind of murder mystery, with a solution that lay millions of years in the past.

9.

NEMESIS

1976–1988

UNTIL THE EARLY SEVENTIES, WALT ALVAREZ RECALLED, his relationship with his father was "not close." He had majored in geology at Carleton College in Minnesota, with financial support from Alfred Loomis, who wanted to spare Walt's father from taking on outside work to cover the cost. Even after Walt started graduate school at Princeton, Alvarez wondered why his son had chosen such an unexciting field. In fact, it was due to Geraldine, who had lent Walt his first rock hammer and taught him to find minerals in the Berkeley hills. Without her, none of what followed would have happened at all.

In 1965, Walt married Mildred Millner, who was known as Milly. After a yearlong scientific honeymoon in Colombia, they moved to the Netherlands, where he worked as an oil geologist, and later to Libya, only to be forced to relocate after Muammar Gaddafi came to power. He ended up at the Lamont-Doherty Geological Observatory at Columbia University, a research center famed for its collection of core samples from the ocean bottom.

At first, Walt remembered wryly, "Dad did not originally think that geology was an interesting science." Despite what the elder Alvarez believed, it was an exhilarating time. After decades of debate, geologists had accepted that the earth's crust was made of slowly moving plates that caused earthquakes, volcanoes, and continental drift. Plate tectonics was clearly visible in seafloor spreading, in which volcanic activity produced new oceanic crust at underwater ridges.

One day, Alvarez unexpectedly proposed a possible way to estimate the age of the oceanic crust at specific locations with the radioisotope

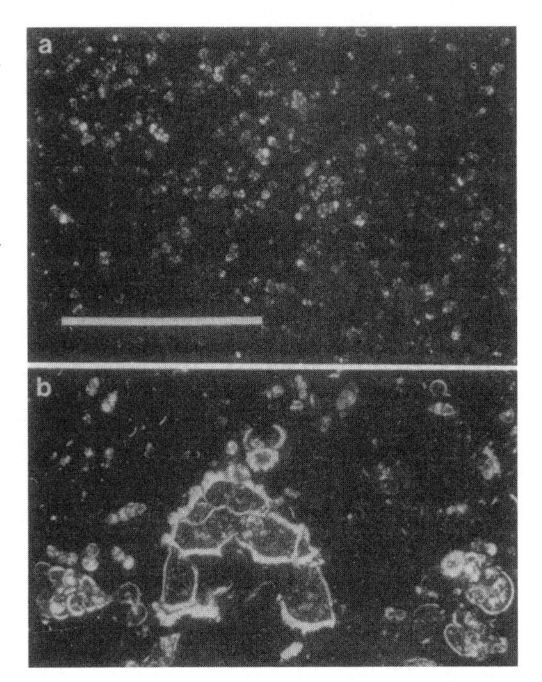

Photomicrographs of layers above and below the K–T boundary at the Bottaccione section at Gubbio, Italy, showing the disappearance of larger forams at the end of the Cretaceous. The American Association for the Advancement of Science.

beryllium-10. After regularly discussing it by telephone with Walt, he decided that it wouldn't work. All the same, it was their first serious attempt to collaborate—Alvarez had directed most of his impulses as a mentor toward his students—and he grew excited by the prospect. "Luis Alvarez freely admitted," Rich Muller said, "that there was nothing he would rather do than write a paper with his son."

Walt was conducting fieldwork at the Apennine Mountains in Italy, near the medieval town of Gubbio. In the Bottaccione Gorge, he saw layers of pelagic limestone, originally from the deep sea, that had been spared from erosion, leaving one of the clearest geological sequences of rock in the world. He began exploring their origins with Bill Lowrie, a specialist in paleomagnetism, which was a valuable tool for reconstructing the movements of the tectonic plates.

When rocks were formed, they sometimes contained mineral grains affected by the earth's magnetic field. Because they initially lined up to point north, they provided a reference point that showed how the crust's orientation had changed. In the early seventies, Walt and Lowrie found many mineral "compasses" at Gubbio pointing in the wrong direction, which supported the theory that the geomagnetic field had reversed repeatedly in the past.

The Bottaccione Gorge was the perfect place to plot a detailed timeline of magnetic reversals. Breaking off a piece of limestone and examining it with a hand lens, Walt could see the microfossils of foraminifera, or forams—unicellular organisms that could be associated with specific geological intervals, allowing him to date the layers. For the related problem of measuring sedimentation rates, Alvarez advised him to contact Muller, hoping that they would hit it off.

In the meantime, Walt was drawn to another mystery. He studied the Gubbio forams with the paleontologist Isabella Premoli Silva, who taught him to identify the borderline between the Cretaceous and Tertiary periods. According to the best estimates then available, the K–T boundary occurred sixty-five million years ago, coinciding with the disappearance of many species, including the dinosaurs.

Walt learned to recognize it on sight. First came a bed of white limestone, filled with abundant large foraminifera—up to a millimeter across—from the end of the Cretaceous. Above it was a layer of dark clay about a centimeter thick. Then came deposits of reddish limestone from the Tertiary, in which the larger forams, once so plentiful, vanished entirely. In the interval recorded in the clay, between two limestone formations, a mass dying had wiped out not just foraminifera, but the most famous extinct animals of them all. And no one knew why.

By the early seventies, the consensus was that the extinction was the result of gradual climate change, perhaps due to continental drift, although Walt had his doubts. Tectonic plates moved at "about the rate at which your fingernails grow," while the disappearance of the forams seemed to reflect a sudden event. The key was the middle layer. It was defined less by the clay itself than by the absence of limestone, which was usually produced by the skeletal fragments of tiny organisms.

If sea life went extinct on a vast scale, halting limestone production, sediment would continue to accumulate from continental erosion, producing the unusual layer—all clay, no limestone—at the K–T boundary. Since Walt was already thinking about deposition rates, it was logical to wonder how long the clay bed had taken to form. The apparent abruptness of the event might be an illusion, if the process had been slower than usual. On the other hand, if it had been deposited in a short period of time, it pointed unmistakably to a cataclysm.

Walt knew that some scientists felt that the Cretaceous extinction had been a sudden catastrophe—a bang, not a whimper. One possible cause, proposed by the astrophysicist Malvin Ruderman, was a nearby supernova. Radiation from an exploding star, as bright as the rest of the galaxy combined, could have destroyed the ozone layer, allowing deadly ultraviolet rays to reach the earth's surface.

Although Walt wasn't sure that a catastrophic scenario was correct, he recognized the question's importance. In language that recalled his father, he later said, "Choosing what problems and what kind of problems to work on is a critical strategic decision for a scientist. The question of the K–T extinction looked like one that could lead in totally new directions."

In 1976, Muller visited Walt at the Lamont-Doherty Geological Observatory in Palisades, New York. The two men had never met in person, and Muller was struck by their first encounter. "For some reason it had never occurred to me that Walter would look like his father. But he did. There he was, tall and lean, with the characteristic smile and blond hair of my former thesis adviser."

On a stroll through the grounds, Walt asked him many questions. "When I told him about the work I had done with his father," Muller wrote, "he became particularly attentive." Muller felt briefly awkward about his status as "a surrogate son to Luie," but he soon saw that Walt wasn't jealous—he just wanted to learn more about how his father thought. Returning to Berkeley, Muller felt that he had found an important new friend, even a brother, and he was eager to contribute to the study of sedimentation rates at Gubbio.

Like Alvarez, Muller was intrigued by the potential uses of beryllium-10 in geology. Be-10 created by cosmic rays in the atmosphere constantly fell into the ocean, where it was incorporated in the sediments that were being deposited at the time. In layers where sediment had been produced at a high rate, there would be less of the isotope compared to material from the seafloor. Conversely, a greater abundance of Be-10 would point to a relatively long and slow formation process.

Looking up the isotope's half-life, Muller saw that it was 2.5 million years, so there would be just enough beryllium in older rocks—including at the K–T boundary—for it to work. After hearing his plans, however, a colleague informed him that the value in the standard tables was wrong. It was 1.5 million years, not 2.5, which would drastically upset the experiment.

When he learned about the discrepancy, Alvarez phoned the man who had measured the half-life. It was none other than Ed McMillan, who dug up his notebooks and confirmed that the published number was a transcription mistake. He also implied that the confusion was somehow Muller's fault: "If you had read my original paper carefully you would have caught the error."

Muller thought that this was unfair, and he was even more disappointed to realize that the beryllium isotope would be thousands of times less abundant than they had expected. There was no way to measure it on the timescales that Walt had in mind, even with accelerator mass spectrometry. Muller reluctantly informed Walt that the project was doomed: "Killed by a misprint."

Alvarez refused to give up. For years, he had been unsure of how to contribute to his son's work, but his excursions into astrophysics had confirmed that his skills could be useful, even in an observational science. Deciding that this was the best way to help Walt, he told himself that the underlying idea was perfectly sound. He just had to step back and properly define what they were trying to do.

All it really required was a substance that was captured in ocean sediment at a constant rate. It didn't need to be radioactive; indeed, it would be better if it didn't decay. As he considered what else might drift down to the seafloor, he recalled that meteors—most no larger than a grain of sand—constantly burned up in the atmosphere, producing thousands of tons of dust each year. From the abundance of this material in a rock sample, he could estimate how long it took to form. He just needed a chemical tracer that could identify meteoritic dust.

Once again, the solution lay in a miscellaneous fact that he had mentally filed away. In Antarctica, he had heard about a technique that might be useful, and a table in the *Encyclopedia Britannica* confirmed his hunch. Elements in the platinum group—including ruthenium, rhodium, palladium, osmium, iridium, and platinum itself—were more common in meteorites than in terrestrial minerals. Billions of years ago, dense liquid iron sank down through molten rock to form the earth's core. Because platinum and other "siderophile" elements alloyed with the hot iron, they were pulled down with it, removing them from the crust.

A meteor, by contrast, was smaller than a planet, so gravity wasn't strong enough for this separation to occur. Given the higher levels of these elements in meteoritic material, their abundance in a layer of sediment would indicate how long it took to form. Looking for a tracer, Alvarez settled on iridium, which could be measured with neutron activation analysis. When a sample was bombarded with neutrons, target atoms were transmuted into radioactive isotopes, releasing characteristic gamma rays that could be counted for each element.

Iridium displayed all the right characteristics. It had a high probability of capturing a neutron, and its radioisotope lasted long enough to be easily measured, while still providing enough decays for an accurate count. On June 20, 1977, Alvarez wrote to Walt about what he had in mind. "There is every reason to believe that the infall of meteoritic material is constant, over long periods of time, so the iridium in that material can act as a tracer, and indicate whether the sedimentation rate of the pelagic limestones is faster or slower than the normal rate."

At that point, Alvarez wasn't focusing on the K–T boundary in particular, and he was advising from a distance. Earlier that year, however, he had spoken to a colleague who wanted to bring Walt to the geology department but was concerned that the available position—an assistant professorship—wasn't senior enough. Alvarez suggested that he try offering it anyway to Walt, who accepted, since it would grant access to the university's exceptional resources. "And I think he was also drawn by his father," Muller wrote, "his fascinating father, whom he knew so little."

In the fall, at thirty-six, Walt moved to Berkeley for the first time since high school. Alvarez naturally wanted to draw him into the world of the lab, and at a Monday night seminar, Walt gave a talk on his research, presenting his father with a gift that would change their lives—a chunk of rock from Gubbio, which he had polished on one side and encased in plastic.

Using a magnifying glass, Alvarez studied the K–T boundary with his own eyes, observing how the large forams below the clay layer disappeared from the limestone above. Examining it up close was very different from just hearing about it, and he thought that it was one of the most exciting things that he had ever seen. He remarked to Walt, "Maybe geology is interesting after all."

It was also a chance to collaborate. As Muller noted of Alvarez, "He had always been somewhat baffled by Walt's interest in geology and in its little problems." When it had just been a matter of sedimentation rates, Alvarez said that Walt could "have the fun" of learning neutron activation analysis himself, but now he decided to take an active role in the more sensational project of solving the Cretaceous extinction, connecting with his son as he never truly had with his own father.

All he needed to complete the picture was Muller, his surrogate son, whom he wanted to pair off with Walt. When asked to participate, however, Muller declined, feeling that Alvarez's desire for one last victory—and his paternal affection—had led him astray: "He had made so many discoveries that he no longer seemed interested in making new ones, unless they had the potential of being truly revolutionary. He was taking big risks, but he was consistently losing."

Alvarez still knew how to assemble the right tools and people. At Berkeley, he had a research reactor at his disposal, as well as Frank Asaro, the chemist who had refined neutron activation into a technique of unmatched precision. Born in 1927, Asaro was a painstaking experimenter—Walt saw him as "the intellectual heir of Tycho Brahe"—and a resourceful scientific detective. He also suffered from such a bad stutter that Alvarez's teenage son, Don, once accidentally hung up the phone on him when Asaro was unable to force out a greeting in time.

Asaro suspected that iridium would be too scarce to detect, but he agreed to participate, mostly out of respect for the persistent Alvarez. The project was delayed by technical issues, limited funding, and a backlog of experiments, but it finally began in 1978. Twelve samples from the Gubbio boundary were exposed to neutrons, then set aside for weeks, allowing other radioisotopes to dissipate. At that point, testing for iridium should have been straightforward, but the result was so surprising that Asaro checked it repeatedly until he was absolutely sure.

When the final numbers were obtained at the end of June, Alvarez immediately called Walt to say that something was "seriously wrong." As Asaro recalled, "We were happy to have found iridium but disturbed to have found so much." There was thirty times as much iridium in the clay as they expected—far more than could be explained by variation in the sedimentation rate. Something else had happened there that had nothing to do with meteoritic dust.

First, however, they needed to confirm that it wasn't just confined to Gubbio. Searching the literature for other potential sites, Walt identified a location in Denmark. At a cliff near Copenhagen, he collected a sample from a thin black clay bed sandwiched between the white chalk from the Cretaceous and Tertiary. When they analyzed it, they found the same iridium anomaly.

As a lifelong "bump hunter," Alvarez was uniquely positioned to seize on the clue. "It's the most exciting kind of science," he later said. "In fact, there's even a word for it: serendipity. I'm sure you know. It means you're not looking for it, and you stumble on it. But you have to be there with your eyes open and recognize that you have something really strange when you see it. So it's not all luck."

He went after the iridium "like a shark smelling blood." Originally, he had seen it as a convenient extraterrestrial tracer that would indicate how long the layer took to form. As it turned out, the method was less than useful—he later learned that other researchers had tried it without success—but he had been handed the key to a greater puzzle. The extinction itself, he saw, had been caused by a killer that came from space, bringing the iridium with it.

Alvarez thought at once of a supernova. An exploding star would release iridium, as well as other telltale elements that might still be in the clay. The obvious tracer was plutonium, which was produced in nature only by supernovas. While most of it would have decayed, one isotope, Pu-244, had a half-life of eighty million years. Finding it in the boundary layer would be conclusive, but far from easy. Although neutron activation could turn plutonium into a detectable isotope, its lifetime was so short that they would need to take readings straight from a radioactive sample.

Since this would pose immense difficulties, Asaro recruited Helen Michel, an expert radiochemist who had worked with him on equally challenging projects. Michel—who, at forty-six, was one of the few women in the field of nuclear chemistry—had started out as Asaro's assistant, but he had quickly promoted her to a position of equal seniority. Most famously, they had used neutron activation to prove that a brass plate held by the Berkeley library, allegedly inscribed by Sir Francis Drake, was a forgery from no earlier than the nineteenth century.

For the task ahead, Michel's skills would be indispensable. First they concentrated the sample in a long series of chemical separations that ended in March 1979. After the clay was irradiated for eight hours, they started to purify it of contaminants. As the countdown began—any plutonium would be decaying rapidly—Asaro and Michel worked all night, with Walt and Milly bringing coffee and cookies. Six hours later, they checked the signal to see what was there.

Shortly afterward, Muller ran into Walt and Alvarez, who said that they were already writing a paper on the supernova hypothesis. He worried that they were moving too fast, but the following day, Alvarez handed him a note at the office: "Meet me in the hallway." Once they were alone, Alvarez whispered the news. "We found the plutonium! Walt and I knew it yesterday, when we talked to you." When

Muller asked why he hadn't said anything earlier, Alvarez explained, "Walt knows that if I don't tell something to you, it means I won't tell it to anyone."

Muller understood why Alvarez wanted to impress his son with the value of secrecy. "It was very rare to have such exclusive knowledge. Maybe there was some important consequence of the supernova theory that we could still find, now that we knew we were right. There was no rush to publish, not yet." They also had to tread carefully, looking for possible errors: "When you have a theory that is verified, that's the most dangerous time in all of physics."

In May, Walt and Alvarez presented their findings at the annual meeting of the American Geophysical Union in Washington, DC. According to press accounts, Alvarez hinted that he had made up his mind about the cause: "A solar flare or a large meteorite striking the earth could also account for the iridium spike, but he indicated that he believed in the supernova theory."

Asaro was less confident. He and Walt consulted Earl Hyde, the lab's deputy director, who said, "Repeat every single step from the very beginning, on a fresh sample, to be absolutely sure there really is plutonium-244 in that clay." It was far from a trivial task, but Asaro was troubled by a memory. At one point in the purification process, he had taken some hydrofluoric acid from a lab upstairs. He later found that the bottle was kept under an exhaust hood that had been exposed to plutonium before. Given the potential for contamination, he decided to rerun the experiment, even though it meant two weeks of hard work.

When they were finished, they found no plutonium in the clay at all. Further investigation confirmed that the Pu-244 had been accidentally introduced into the sample in the lab. It wasn't a smoking gun, but a mistake. Although Alvarez was disappointed, he was relieved that they had avoided an embarrassing retraction. For a scientist who obsessively looked for "trouble," it had been a rare oversight. "There are several people in the world to whom I give credit for having saved my life," he told Muller. "In that list I now include Frank and Helen."

Alvarez was back where he started, but he refused to give up. "He told everyone who was interested that he was trying to solve the most exciting problem of his life," Muller said, "but I don't think many people took him seriously, not even Frank, Helen, or Walt." Since he had been obliged to publicly walk back the supernova hypothesis, he needed to find an alternative. For a month and a half, he came up with new ideas every week, generating them in his meditation sessions at home before sharing them at the lab for everyone to debate.

He reminded himself that the iridium wasn't the important part. It was only a tracer that pointed to space, so it might have been accompanied by something else. One logical candidate was hydrogen, the most abundant element in the universe. Perhaps, he thought, it had reacted in the atmosphere to make water vapor, locking up oxygen, which would asphyxiate dinosaurs and other organisms. Alvarez ticked off possible sources—the sun, a giant molecular cloud, an asteroid strike on Jupiter—but couldn't get the hydrogen idea to work.

Another tempting suspect was an asteroid colliding with the earth. Christopher McKee, an astronomer at Berkeley, suggested that an ocean strike would produce a monster tsunami, but Alvarez couldn't imagine how a localized collision would affect marine life or inland regions, let alone every dinosaur on the planet. A third possibility was a near miss with an asteroid that disintegrated in the atmosphere, creating a dust cloud that was spread by the earth's rotation. The darkness would halt photosynthesis, killing off plants and larger animals, but Alvarez decided that the atmospheric drag wouldn't break an asteroid apart.

All the same, he was moving in the right direction. Brooding over the effect of an impact, he focused on the subset of asteroids that intersected with the earth's orbit. Based on the numbers and sizes of these "Apollo objects," an asteroid three miles in diameter could be expected to strike once in a hundred million years. With a calculator, he found that it would yield an explosive force of one hundred million megatons of TNT—over a billion times greater than Hiroshima. This would be fatal for anything directly beneath it, but it wasn't obvious how it would cause a mass extinction.

What he needed was a killing mechanism. At last, he remembered the dust that he had considered for the flyby scenario. An asteroid that struck the earth directly would vaporize the underlying rock, ejecting debris into the atmosphere. If the cloud rose high enough to be spread across the planet by winds, it could block out the sun. It reminded him of the dust from nuclear bomb tests, and—even more crucially—of the eruption of Krakatoa. The volcanic island between Sumatra and Java had exploded in 1883, spewing enough rock and ash into the air to produce cooling and unusual weather patterns around the world.

Decades earlier, his father had presented him with a copy of *The Eruption of Krakatoa*, a handsome volume published by the Royal Society of London, which Alvarez had given to Walt. Since his son was currently in Italy, Alvarez asked someone to retrieve the book from Walt's house. Leafing through it, he came across a discussion by the meteorologist E. Douglas Archibald of the ensuing "twilight glows, coloured suns, and cloud-haze." The effects were observed until 1886, leading Archibald to conclude that the dust spouted "into the loftier regions of the atmosphere would be suspended there for *at least two years*."

This estimate later turned out to be wrong. In reality, the dust largely dispersed within a few months, leaving trace amounts that produced spectacular sunsets, but little else. The error led Alvarez to drastically overstate how long the darkness produced by an asteroid might last, but for now, it didn't occur to him to question it. With Walt away in Europe, Alvarez tried the idea on the others in July, and it held up surprisingly well. "After two weeks with no change in the model," Muller wrote, "I began to wonder if Luie had really found the solution."

In September, Walt planned to attend a conference on the K–T extinction in Copenhagen. Assuming that the participants "would be delighted" by his theory, Alvarez wanted to join, but Walt was more cautious: "I knew geologists and paleontologists better than Dad did, and I was pretty sure there would be strong resistance, and even hostility, to a nonuniformitarian explanation." The conventional wisdom was that past events occurred at the same rate as in the present, slowly and gradually, not through chance catastrophes. Geologists resisted sudden change, both in hypotheses and in their own field, and the impact scenario would be furiously contested.

Alvarez, who was less concerned by these issues, began drafting a paper with Walt, Asaro, and Michel, leaving one longtime associate on the outside. For years, Muller had been Alvarez's closest professional confidant, at least "until the son arrived," and he began to suspect that he had underestimated his mentor. "Now it was too late. Had I made slightly more effort I might have come up with some ideas, or have helped with the measurements. I was sure that Luie then would have welcomed my coauthorship of this paper. Here was one of the greatest scientific papers of the century, and I was just a bystander."

AS THEY PREPARED THE PAPER for publication, Alvarez and his colleagues knew that it might become a classic, but it was even more certain that it would be challenged by critics. To methodically make their case, they described iridium spikes in Italy, Denmark, and a recently discovered third site in New Zealand, which indicated that the anomaly was indeed worldwide. A separate set of tests had established that the clay at the K–T boundary differed in chemical composition from neighboring regions, pointing to an extraterrestrial source.

The most plausible cause, they wrote, was the impact of an asteroid about six miles in diameter, an estimate that they based on the total amount of iridium and clay that the collision had spread across the earth. All the pieces seemed to fit, with one conspicuous exception—there was no known crater of the right size and age. Since it seemed probable that the asteroid had struck the ocean, which covered two-thirds of the earth's surface, the group concluded that the site might never be found.

Alvarez felt confident enough to circulate a draft. One recipient was Malvin Ruderman, an advocate of the supernova hypothesis, who sent a congratulatory note: "You are right and I am wrong." Less convinced was Phil Abelson, Alvarez's former graduate student, who had once narrowly missed discovering nuclear fission. Now he was the editor of *Science*, the world's leading scientific journal, which would be the best home for a major paper. Over dinner with Walt, however, Abelson said that he had recently printed several competing theories about the extinction of the dinosaurs: "And at least N-1 of them must be wrong."

Eventually, Abelson accepted a shortened version, which received a rigorous review. One reader was David M. Raup, a paleontologist at the University of Chicago, who later speculated that such "maverick ideas" could be published only by someone of Alvarez's stature. Previous impact proposals had been made by the scientists Digby McLaren and Harold Urey, each "at the height of his career and influence." As Raup noted, "Perhaps crazy ideas continually sprout in the minds of some small fraction of scientists in all age groups, but it is only the leading people who have the personal confidence to try them out and clout to get their papers accepted."

The team began to strategically spread the word. On January 4, 1980, at a meeting of the American Association for the Advancement of Science in San Francisco, Alvarez described how the dust could block sunlight: "The resulting global darkness would suppress photosynthesis for a period of at least three to five years." Larger animals with low populations, like dinosaurs, would be especially vulnerable, while their smaller competitors—including the mammals that were fortuitously positioned to take their place—could survive "on nuts, seed, insects, and decaying vegetation."

Alvarez's talk and a preprint of the article produced what he called "a humming telephone and letter network all across this country," culminating in the publication of "Extraterrestrial Cause for the Cretaceous-Tertiary Extinction" in the June 6 issue of *Science*. Rather than the enthusiasm that he had expected, however, the initial response from dinosaur experts was overwhelmingly negative. Raup found that many scientists saw Alvarez and his collaborators as outsiders who lacked expertise in paleontology and—even more damningly—had been "too quick to call in the press."

Helen Michel, Frank Asaro, Walter S. Alvarez, and Luis W. Alvarez on November 19, 1980. Lawrence Berkeley National Laboratory.

"And so was born one of the more stunning hypotheses of modern science," the *New York Times* journalist John Noble Wilford later wrote. As news articles proliferated, critics raised legitimate concerns, including the apparent lack of any effect on plants. In response, Alvarez argued that plant species could survive a period of darkness—eventually scaled down to three months—as seeds sprouted after sunlight had returned. He received additional support from a group led by Carl Orth at the U.S. Geological Survey in New Mexico, who found a sharp decline in fossil pollen near the iridium anomaly. Plants had suffered as much as the animals.

Other skeptics thought that the iridium had precipitated out of seawater. When Alvarez's team sought to search an area that had always been on land, however, they were denied a grant, reportedly because the reviewer thought that "we would be wasting our time and the government's money . . . because the iridium certainly came out of the ocean." Later, an apparent iridium anomaly in Montana turned out to be an error caused by a technician's platinum ring, which had briefly rubbed against the sample. In any case, Orth's iridium was from a continental deposit, indicating that the chemical precipitation hypothesis was incorrect.

Another objection came from the Berkeley paleontologist William Clemens, who furnished samples from the Montana boundary layer. After Asaro and Michel found iridium there, Clemens declared that it was three meters above the last dinosaur fossil, implying that the extinction had predated the impact by thirty thousand years. Alvarez preferred to say that the gap was "just three meters." Because dinosaurs were rare—their remains had an average vertical spacing of one meter—a fossil was unlikely to appear at exactly the right depth.

Concluding that the discrepancy was a sampling issue, Alvarez proposed using a sonar technique to conduct a more thorough search. It was never implemented, and Clemens maintained that an asteroid didn't fit what paleontologists saw in "the evolutionary history of the group they studied." As he and his coauthors wrote in a rebuttal paper, "Different groups drop out of the fossil record at different stratigraphic levels." This pointed to a slow extinction over many thousands of years, with numerous species vanishing long before the alleged collision.

To reconcile these two scenarios, some scientists proposed that the impact wiped out species that were already in decline. One early advocate of the Alvarez hypothesis was Stephen Jay Gould, the Harvard evolutionary biologist and acclaimed science writer, who suggested in a letter that "the biosphere was already weakened for other reasons when your asteroid hit." Alvarez was unpersuaded, and he was backed by the paleontologists Phil Signor and Jere Lipps of the University of California at Davis, who used an elegant statistical argument to show that a seemingly gradual extinction might be an illusion.

Signor and Lipps invited readers to imagine a diverse ecosystem that suffered a sudden catastrophe. Paleontologists could estimate when each species died out by the location of the last known fossil, which was unlikely to literally be the final surviving individual. More of these "last occurrences" would naturally be found at levels closer to the time of the actual cataclysm, with the number decreasing in layers from further back in the past. This created the false impression—which would diminish as more remains were uncovered—that the extinction rate had gradually risen.

As Alvarez and his associates tried to anticipate their critics, they were still operating without formal support—the iridium analyses had been squeezed in between other experiments, and they were being underwritten for now by a general development fund. Their immediate goal was to find iridium elsewhere, which was a key prediction of the theory, but they were impeded by the slow pace of neutron activation. Chemically preparing the samples took a whole weekend, resulting in a bottleneck that limited how many sites they could test.

After talking with Walt nightly by phone for months, Alvarez found a solution. Radioactive iridium often emitted two gamma rays, and by distinguishing both—rather than just one—from background radiation, they could bypass the laborious purification stage. His new detector, the iridium coincidence spectrometer, would be capable of automatically processing thousands of samples every year, like a version of his data system for the bubble chamber. Before long, iridium anomalies at the K–T boundary were seen in over eighty locations across the world.

Other clues appeared in the clay. The Dutch geologist Jan Smit found glassy droplets that were identified as microtektites—tiny spheres that could have been made by a splash of molten rock. Most convincing of all were bands of "shocked quartz." To supporters, these deformed crystals, which were also seen at nuclear test sites and meteorite craters, clearly reflected an impact. Alvarez felt that the triple coincidence of iridium, microtektites, and shocked quartz was conclusive, even if the exact mechanism had yet to be determined: "Scientists have known there was a murder and we feel we have found the killer. But we're still not sure what weapon was used."

Over the summer of 1981, Alvarez visited Gubbio with Janet and the children. In the gorge, Alvarez asked why the layers were inclined at an angle, rather than level with the ground. Walt replied that the answer was "complicated," prompting his father to preemptively neutralize a possible point of confusion: "He leaned over and had [his] picture taken with the camera tilted, so that audiences of physicists would understand the originally horizontal beds."

Frank Asaro—who diligently overcame his stutter until he could give public talks—never forgot Alvarez's insistence on the importance of shaping the narrative. For Alvarez, debate was a contact sport, and he claimed not to understand why paleontologists had to be "dragged, kicking and screaming," into acceptance. In his less generous moments, he lumped them together with the "crazies" whom he encountered in the Kennedy investigation, and he liked to quote the biologist Peter Medawar: "Many scientists unconsciously deplore the resolution of mysteries they have grown up with and have therefore come to love."

Walt was more sympathetic. Although his younger brother, Don, saw them as playing "good cop, bad cop," Walt knew that paleontologists had been trained in the uniformitarian philosophy of the geologist Charles Lyell, who had fought back against catastrophic scenarios based on the myth of the Biblical flood. The dinosaur extinction was exactly the kind of problem that attracted cranks. As Muller noted, the Alvarez hypothesis "had all the characteristics of a nut theory."

If nothing else, their resistance gave Alvarez the advantage of a head start, allowing him to refine his ideas without any competition. "And yet in the fifteen years that I had known him, I had never seen him so unhappy," Muller said. "He could simply wait for history to make the judgment and prove him right; yet he took a great deal of time and trouble to respond to all the papers criticizing his theory. He sometimes was so furious at some of the papers and letters attacking his model that his complexion noticeably reddened." The next generation might be more receptive, but Alvarez couldn't afford to be patient.

A handful of scientists were more favorably predisposed. David Raup had long been fascinated by whether extinction had more to do with "bad genes or bad luck," while Stephen Jay Gould admitted his own "idiosyncratic preference for rapidity." Gould was best known for his evolutionary theory of punctuated equilibrium, in which long dormant periods alternated with intervals of dramatic change. As he later wrote of his early support of Alvarez, "Always look to reasons of personal interest, rather than general wisdom, in such cases."

The opposition tended to come from the specialists, who constituted the majority of the field. William Clemens, who worked out of the same building as Walt, continued to describe the theory as "codswallop." Along with his other arguments, Clemens said that fossil evidence of mammals from the Paleocene appeared below the clay layer, implying that they evolved before—not after—the impact. Jan Smit countered that the channel that buried the fossils cut through the K–T boundary itself, so the river bottom contained remains from both before and after the iridium anomaly.

Clemens met with the group every Tuesday, and while their interactions were always cordial—Walt called him "a real gentleman"—he refused to concede. He saw more ammunition in a section of the Montana clay layer that seemed to have been laid down at a different time than the rest. After Walt pushed back, a review of the discrepancy revealed that it had been a transcription error. As Muller pointed out, "It is rare for a theory to be so good that it can be used to predict that certain measurements have been done incorrectly." Clemens remained unconvinced, prompting a crack from Alvarez: "Bill's theory is that our theory is wrong."

Alvarez was less patient with paleontologists whom he encountered only in passing. A 1981 extinction conference in Ottawa was the scene of an acrimonious encounter with Dewey McLean, who favored a mechanism involving the Deccan Traps, a huge volcanic feature in western India that dated from the end of the Cretaceous. Violent eruptions over thousands of years, he said, could have released carbon dioxide, leading to a devastating greenhouse effect.

As McLean spoke, he recalled, Alvarez "glared red-faced at me across the tables that separated us." When the group broke for coffee, Alvarez drew him into a corner. "Do you plan to publicly oppose our asteroid?"

"Dr. Alvarez, I've been working on the K–T for a long time," McLean responded. "I published my greenhouse theory two years before you published your asteroid theory." He wasn't dissuaded when Alvarez told him what had happened to Buford Price, and he linked the dispute to the debate over climate change. "We're facing a possible greenhouse today. I have an obligation to continue my work."

"You've been warned," Alvarez said flatly. Then he turned away to join the other scientists without looking back.

Despite its melodramatic tone, the story was plausible—McLean had no other reason to know Buford Price's name—and Walt later confirmed their "heated exchanges." In the aftermath, McLean alleged that Alvarez started "a widespread virtual vendetta" against him, derailing his career at Virginia Polytechnic Institute and causing him to suffer crippling joint pain from stress. Alvarez denied this, but in a letter to Robert Jastrow, another skeptic, he wrote:

> As a horrible example of what can happen to a scientist when he clearly—in sight of all his peers—ignores important evidence, I'll remind you of Dewey McLean, a paleontologist from V.P.I. Dewey used to be invited to all the conferences and debates about the K–T extinction. But he no longer is, because he refuses to face up to the fact that the [iridium] enhancement is of great importance. . . . So Dewey is now a forgotten person in the field, or when he is remembered, it is only for a few good laughs, at the cocktail party at the end of the Deweyless meeting.

McLean suspected that he knew why. Shortly after the *Science* paper appeared, Alvarez was invited to a NASA conference in Massachusetts to discuss what could be done if another Apollo object was found on a collision course with the earth. He later participated in a study that concluded that instead of simply blowing it up—since the fragments could be deadly in themselves—it would be best to land on the asteroid, drill a hole, and set off an explosion inside, nudging it into a safe orbit.

Alvarez was sure that such a threat would occur sooner or later: "All I can say is we'll be one day closer tomorrow." As a first step, the NASA advisory council proposed the Spacewatch Project, an initiative to track all asteroids greater than thirty feet in diameter. McLean wrote acidly that the prospect of "death from the heavens—imminent, unavoidable, catastrophic, horrifying mass extinctions of life on earth—was just what the space agencies needed to rejuvenate them via new funding and career opportunities." In reality, Alvarez and his allies would have seen the program as a way to build support for his theory, not the other way around.

Since the debate was inherently interdisciplinary, a conference was held in October 1981 at Snowbird, Utah, to encourage a free exchange of ideas. Groups from Los Alamos and Pasadena presented computer models—originally designed for hydrodynamic studies of nuclear weapons—that showed how dust could reach both hemispheres. Debris would be sent skyward at ballistic velocities, projected up the evacuated column punched in the atmosphere by the asteroid.

Another question was whether global temperatures would fall or rise. The dust would block the sun, but it could also absorb and radiate heat, while vapor from an ocean impact might produce a greenhouse effect. Brian Toon, a NASA physicist from Ames Research Center in California, delivered the results of a model originally built by the astrophysicist James B. Pollack to simulate storms on Mars. Toon said that sunlight at the earth's surface would fall by half, cooling it dramatically.

In the audience were William Moran and Lee Hunt of the National Research Council, who proposed that a similar computer program be used to model the climactic effects of a nuclear war. When it accounted for the effects of dust from bombs and smoke from firestorms, the results were horrifying. Even a limited exchange would lower temperatures below freezing, leading to the hypothetical ecological disaster that became known as nuclear winter.

Pollack and Toon were both former students of the astronomer Carl Sagan, who had recently hosted the blockbuster television series *Cosmos*. Along with advising on the nuclear winter model, Sagan began to sound an urgent warning. In the second sentence of a landmark paper in *Science*, Sagan and his collaborators wrote, "The discovery that dense clouds of soil particles may have played a major role in past mass extinctions of life on Earth has encouraged the reconsideration of nuclear war effects." Elsewhere, Sagan described the potential effects of famine and darkness: "There seems to be a real possibility of the extinction of the human species."

On publication, the nuclear winter scenario was challenged at once, with reputable scientists on both sides. Alvarez himself wasn't sure whether he believed it. Years earlier, he had told a correspondent that dust in a nuclear war would be "*almost* the last thing that anyone should worry about." Eventually, he came to see the theory as "seriously wrong," but he didn't want it to reflect badly on the asteroid hypothesis, so he kept his concerns to himself.

As usual, Alvarez was willing to use an unverifiable but plausible idea to advance his own objectives. At Bohemian Grove, he convinced Barry Goldwater that the threat was worth taking seriously, hoping to discourage politicians from thinking that such a war was winnable. In his memoirs, he said that the most attractive aspect of the hypothesis "is that no one has been able to disprove it," and he concluded, "It may turn out that my rather peculiar way of thinking about new problems will have had an important effect on what I continue to believe is the world's number one problem, the avoidance of nuclear war."

Alvarez believed that the underlying question had been settled on its merits. In a talk at the National Academy of Sciences, he said, "That the asteroid hit, and that the impact triggered the extinction of much of the life in the sea, are no longer debatable points." He told his audience that physicists could react "instantaneously" to evidence that disproved their theories, although he failed to mention the many contrary examples—from Buford Price onward—from his own experience. "But that is not true in all branches of science, as I am finding out."

He was astonished that scientists could seriously believe that the dinosaurs vanished for no particular reason shortly before the most devastating catastrophe in history. Noting that he was standing in front of a projection screen that was twenty feet wide, he invited his listeners to imagine that it represented the nearly two hundred million years that dinosaurs ruled the earth. He then asked them to guess how much space would separate the supposed earlier extinction and the K–T event. Holding up a white card, he showed them that the lines were less than a millimeter apart.

In 1984, a poll of hundreds of paleontologists, geophysicists, and geologists indicated that only a fifth endorsed the Alvarez hypothesis. A greater number doubted the impact altogether, and a stunning twelve percent didn't think that a mass extinction had occurred at all. Robert T. Bakker, perhaps the most celebrated dinosaur paleontologist alive, later criticized the "arrogance" of the catastrophists: "They know next to nothing about how real animals evolve, live, and become extinct. But despite their ignorance, the geochemists feel that all you have to do is crank up some fancy machine and you've revolutionized science."

A separate conflict persisted over volcanism. Walt sometimes wondered if an asteroid strike could have caused or intensified the eruptions, but the scenarios were usually seen as mutually exclusive. Alan Rice, a physicist at the University of Colorado, complained that Alvarez told him "to sit down" when he raised the issue after a lecture. Charles Officer and Charles Drake of Dartmouth, the loudest voices in favor of the volcano theory, argued that iridium and other elements in the clay had been deposited over thousands of years by the Deccan Traps in India.

The controversy reached its peak with an unsigned editorial in the *New York Times* on April 2, 1985. In "Miscasting the Dinosaur's Horoscope," the paper attacked the Alvarez hypothesis, stating—as if the argument were settled—that the dinosaurs had been declining long beforehand and that the iridium came from volcanoes. "Terrestrial events, like volcanic activity or changes in climate or sea level, are the most immediate possible causes of mass extinctions. Astronomers should leave to astrologers the task of seeking the cause of earthly events in the stars."

Gould was shocked that the paper "would editorialize against a theory so clearly subject to empirical test and so eminently interesting as well." In a letter to the editor, Walt and Muller said that they would have thought it was "an April Fool's joke" if it had been printed one day earlier. Muller was later told that the editorial department hadn't consulted the science desk before publishing the piece, the tone of which inspired "concern among the funding agencies in Washington."

A month later, a considerably more positive story appeared in *Time*. On the cover of the May 6 issue, a green Tyrannosaurus looked back at a mushroom cloud raised by a distant impact. The article said that the Alvarez hypothesis was now accepted by "all but a few diehards," and it implied that scientists were moving on to an even more astonishing idea. It came from Muller, who had finally made his mark on the debate with a theory that encompassed not only the K–T event but other mass extinctions in the past, and perhaps even millions of years into the future.

IN ONE OF HIS EARLIEST LETTERS on the subject, Alvarez had mentioned looking for a killing mechanism "probable enough that we could use it to explain earlier extinctions." He and his colleagues sounded the same note in their *Science* paper, writing that if their theory was correct for the K–T event, "the same could be true of the earlier major extinctions as well." Any argument along those lines would center on the canonical "big five" extinction events, ranging from the late Ordovician—445 million years ago—to the end of the Cretaceous.

One avenue of approach was to search for iridium anomalies in other strata that contained the glass droplets identified as microtektites. Walt suggested looking at the Eocene-Oligocene boundary, in which a comparatively minor extinction occurred thirty-three million years in the past. As they had hoped, Asaro and Michel saw an iridium spike at that layer—the second level where one had been found. Alvarez felt confident that another asteroid strike was responsible.

His speculations were upended by an even more astounding claim. At the University of Chicago, David Raup had teamed up with the paleontologist Jack Sepkoski, who compiled a database of the first and last fossil occurrences of 3,500 taxonomic groups. When they examined printouts of extinction trends, they saw what looked like a cyclical pattern. It was clearly visible when they looked at the charts from across the room, and a more rigorous analysis revealed twelve extinction peaks that seemed to occur every twenty-six million years.

In the fall of 1983, they circulated a draft paper on the apparent periodicity. On hearing about it, Alvarez was wary of endangering his own theory's credibility by endorsing such a wild idea. He knew how easy it could be to find nonexistent patterns in noisy data—like the Dallas police recordings—and thought that Raup and Sepkoski should have omitted the Cretaceous and Eocene extinctions, since impact events couldn't possibly be periodic.

When Muller questioned this last point, Alvarez said that keeping the Cretaceous and Eocene events would be "cheating." Feeling that this showed a lack of imagination, Muller asked, "Suppose someday we found a way to make an asteroid hit the earth every 26 million years?"

Alvarez demanded a mechanism. "How could asteroids hit the earth periodically?"

Although he disliked being put on the spot, Muller quickly thought of a potential answer. "Suppose there is a companion star that orbits the sun. Every 26 million years it comes close to the earth and does something. I'm not sure what, but it makes asteroids hit the earth."

To his surprise, Alvarez took it seriously, even though Muller had spoken on the spur of the moment. "I think that your orbit would be too big. The companion would be pulled away by the gravity of other nearby stars."

Going to the chalkboard, Muller came up with an orbital diameter—based on twenty-six million years per revolution—of just under three light years, which seemed small enough to be stable. Alvarez was impressed. "Let's call Raup and Sepkoski and tell them that you found a model that explains their data."

When Alvarez called Raup, he found that a similar proposal had been shot down just one week earlier. Gene Shoemaker—a geologist whose interest in craters led him to become a leading authority on the dynamics of comets and asteroids—was skeptical of the idea of a companion star, concluding that the orbit would be unstable if it passed close enough to affect the asteroid belt. Alvarez agreed, but he told

Raup, "I really hope that your analysis is correct, because it would give everyone something very puzzling to explain."

In December, Muller saw a *New York Times* article on Raup and Sepkoski's results, which told him that he had blown his lead—countless scientists would be looking into it now. "A few months earlier I had assumed that the giant puzzle of the dinosaur extinctions had been solved completely by Luie and his team, with no role for me to play. Now, in effect, I had gradually come to believe that one edge of the jigsaw puzzle wasn't really an edge, but had a few pieces sticking out."

Two weeks later, Muller rode his motorcycle to campus to see Marc Davis, an astronomer with whom he had discussed the problem. They described it to Piet Hut, a Dutch expert in orbital dynamics, who seemed fascinated even by ideas that had been proved wrong. "To be a really good theorist," Muller later said, "you must find even *failed* theories interesting, for someday you may have to apply them in a new situation, where they might prove useful."

Davis had suggested a different model, involving the revolution of the solar system in the Milky Way. As the sun traced its slow orbit, it went up and down relative to the galactic plane, like a horse on a carousel, entering the densest region every thirty-three million years. It was tantalizingly close to Raup and Sepkoski's cycle, but there was no plausible killing mechanism, and the timing was off—the sun was much closer to the galactic plane at the moment than the hypothesis required.

Hut also felt that a companion star would be too distant to affect the asteroid belt. "Of course, the star would have the same effect on the comet cloud. The star would scatter some of the comets toward the earth, just as it scattered the asteroids. Could the impact that killed the dinosaurs have been a comet?"

It hadn't occurred to Muller or Davis. In the fifties, Jan Oort, another Dutch astronomer, had proposed the existence of a distant region—over half a light year from the sun—filled with billions of comets. A companion star could pass periodically through the Oort cloud, sending a thousand additional comets toward the earth every year. The timing implied that the star was near its maximum distance now, and if it was a dim red dwarf, astronomers could easily have missed it.

Muller was excited, but when he tracked Alvarez down in the cafeteria, he was disappointed by his subdued reaction: "I had the feeling he was treating me politely, as if he didn't want to interfere with my enthusiasm." Determined to make his case, Muller began writing a paper about his hypothetical companion star, which he called Nemesis, after the Greek personification of divine justice.

In early 1984, Walt—who was more receptive than his father—showed Muller a map of impact craters. There were about ninety in all, mostly in North America and Europe. After Muller was unable to guess why, Walt explained, "Because that is where the most geologists live."

When Walt plotted a timeline of craters on graph paper, throwing out the ones without reliable dates, he was left with around two dozen, and the gap between the largest craters seemed to fit Raup and Sepkoski's periodicity. After conducting a computer analysis with Saul Perlmutter, one of Muller's graduate students, they found a strong peak at 28.4 million years.

While they were finishing up, Alvarez came into the office. On hearing the results, he wasn't impressed. Although they had calculated that the odds of the pattern arising by chance were less than one in a hundred, it wasn't enough for such an extraordinary claim. Noting that even a small disparity in the timing would cause the extinction and impact cycles to fall out of phase, he concluded, "If you publish this, Rich, they will just laugh at you."

Shortly afterward, Alvarez returned with a pair of articles about the "kappa meson," a resonance particle mistakenly identified decades earlier through bumps in the histograms. It had passed all the statistical tests, he said, but was never seen again: "This was the only particle 'discovered' by my group that ever turned out to be completely wrong." Taping the papers to the door, he left them as a reminder that even good scientists could deceive themselves.

Even after they resolved the timing discrepancy, Alvarez remained unconvinced, until Muller came up with an ingenious plan to persuade him. During the golden age of the bubble chamber, Alvarez had developed a computer program, GAME, to test whether a bump was meaningful. Generating a pile of random histograms, he would shuffle them with real data and ask physicists which ones were most significant. In many cases, they were fooled by the simulated charts.

Perlmutter and Muller prepared a similar test. Handing a stack of twenty plots to Alvarez, they told him that they were simulations of crater dates. Alvarez went through them all in ten minutes, pulling out any that seemed to show periodicity. Holding up his selections, he announced, "This one is the best, and these two are tied for second place. And the second one here I recognize as the real data. I knew you would stick that one in, Rich."

Muller confirmed that Alvarez was right—but then he revealed the twist. Along with the actual numbers, he had included three imaginary plots that simulated how the data would look if crater dates really were periodic. Alvarez had chosen two of them. "Luie hadn't picked any random data at all," Muller wrote. "In making his choices, Luie had actually shown that our data were as good as you could expect a truly periodic signal to be."

Alvarez remarked, "That was very clever." Although he said only that he would look at the problem again, he was coming around to the notion of periodic extinctions. If they were right, the payoff would be so enormous—worthy of a Nobel Prize—that it made up for the remote odds of success, and he was pleased that Walt and Muller were collaborating: "I was delighted to watch two of my closest friends working together for the first time."

Instead of leading, he reinvented himself as a supporting player, looking for ways to be useful, holding everyone to a high standard, and taking over only if needed. The obvious next step was to find Nemesis itself. Perlmutter ran a computer search for candidates in two standard star catalogs, and when a third tape refused to load, Alvarez painstakingly eliminated nine thousand entries by hand.

One day, Alvarez approached Muller with another idea. "Nemesis is right now about as far away as it can get, and it is just turning around, beginning its fall back toward the sun and the earth. That means it is one of the very few stars in the sky that is not moving toward or away from us. We can find Nemesis by looking for a star with nearly no red shift."

As Muller knew well, light from stars moving relative to the observer exhibited a visible change in wavelength. Alvarez suggested using a telescope filter that would register only light from stars where this shift was imperceptible. Muller wasn't sure that it would work—the parameters would be hard to get right—but realized that he should also be searching for Nemesis. "The real effect of Luie's invention was to wake me up. I had already begun to be lazy again."

In the meantime, Alvarez's team looked for iridium spikes at other extinction events, with results that were provocative but inconclusive. Elsewhere, the core of the Nemesis hypothesis—Raup and Sepkoski's periodicity data—was scrutinized by skeptics, who felt that the fossil record was too messy to reveal meaningful patterns. Raup thought that any noise in the data should make it look more random, not less: "To exaggerate only a little, if periodicity shows through in spite of uncertain taxonomy and geologic dating, it must be there!"

At a Berkeley conference organized by Alvarez, the astrophysicist Daniel Whitmire proposed that the cycle was caused by a tenth planet—"Planet X"—moving through a comet disk outside the orbit of Neptune. Gene Shoemaker remained doubtful of Nemesis, noting that no binary stars like it had been observed. Alvarez argued that it was possible simply because they were there to discuss it: "Perhaps periodic extinctions are necessary to give complex life forms, such as humans, a chance to unseat the dominant and primitive species that occupy all the ecological niches."

All the while, Muller was wondering how Nemesis might be found. Since it would be closer to the sun than any other star, it would exhibit correspondingly greater parallax—its apparent change of position caused by the earth's movement. If they took thousands of photos at different times, they could compare them with a computer, looking for a point of light that jumped back and forth. With a start, he realized that he already had everything he needed. His automated supernova search, which was still ongoing, could be repurposed for Nemesis.

Alvarez called it "the most exciting idea he had ever heard of." It was ultimately implemented at a telescope in Lafayette, California, starting with a few thousand candidates from star catalogs before expanding to the rest of the sky. Alvarez was confident that they would eventually discover Nemesis, which he saw as the culmination of a career spent "finding needles in haystacks."

AS HE NEARED HIS LATE SEVENTIES, Alvarez seemed to be streamlining his life. For years, he had enjoyed impulsively flying colleagues to lunch in Monterey in his Cessna 310: "I found few activities as satisfying as being pilot in command with responsibility for my passengers' lives." When he finally sold the plane, he replaced it with another toy. After his patent attorney warned him about the condition of the brakes on his Ford Mustang, Alvarez brought his daughter Helen to the auto dealership. On her recommendation, he bought a red Porsche 911.

Still looking for the elusive invention that would guarantee his financial security, he designed a stabilized binocular telescope for the optics company that he ran with Janet, while a former student, W. Peter Trower, assisted with a "nitrogen camera" to detect explosives in luggage. He continued to serve on committees, including the National Commission on Space, where he joined Neil Armstrong and Chuck Yeager in calling for the colonization of Mars within three decades.

The last piece in his legacy was an autobiography. Although Trower produced a manuscript from thousands of pages of dictation, an executive at the Alfred P. Sloan Foundation, which provided financial support, described it as "unpublishable." To save it, they hired Richard Rhodes, the author of *The Making of the Atomic Bomb*, a future Pulitzer winner that Alvarez had read in galleys. Rhodes cut the book, reworked its structure, and added material on the dispute with Oppenheimer, which Alvarez had neglected to cover in the first draft of *Adventures of a Physicist*. When asked why, he said, "I don't want to go there."

Elsewhere, he was eager to pass along his wisdom. In the fall of 1986, he sat down in the cafeteria across from a postdoc, Cliff Stoll, who said that he was working as a system administrator at the laboratory computer center. After noticing a tiny accounting error, Stoll realized that an unauthorized user was searching the network for information on defense projects. He wanted to track the hacker, but felt frustrated by a lack of support from the lab.

Alvarez replied with a lifetime's worth of experience. "Don't try to be a cop, be a scientist. Research the connections, the techniques, the holes. Apply physical principles. Find new methods to solve problems. Compile statistics, publish your results, and only trust what you can prove. But don't exclude improbable solutions—keep your mind open."

Stoll was still worried about the practical side. "But who's going to pay my salary?"

Alvarez said that he could try writing a proposal that would probably be ignored. "Or you could just chase the bastard. Run faster than him. Faster than the lab's management. Don't wait for someone else, do it yourself. Keep your boss happy, but don't let him tie you down. Don't give them a standing target."

Stoll recalled that Alvarez's advice was crucial: "He'd outlined how to scientifically research a hacker." As recounted in his best-selling book, *The Cuckoo's Egg,* Stoll began to log the intruder's online movements, building a profile as Alvarez might have reconstructed a particle using data from the bubble chamber. The trail eventually led to Markus Hess, a West German hacker who was selling secrets to the Soviets. Stoll credited the lunchtime conversation with setting him on a path that few had traveled: "From that grew the beginnings of computer security."

Any sense of tranquility in Alvarez's final years was disrupted by another piece in the *New York Times.* On January 19, 1988, it published an article by Malcolm W. Browne titled "The Debate over Dinosaur Extinctions Takes an Unusually Rancorous Turn." In a telephone interview, Alvarez did nothing to correct this impression: "I don't like to say bad things about paleontologists, but they're really not very good scientists. They're more like stamp collectors."

He was paraphrasing a quotation attributed to his hero Lord Rutherford: "All science is either physics or stamp collecting." Taken in isolation, however, it sounded arrogant, and other quotes were equally inflammatory. Alvarez reportedly called Bill Clemens "inept" at reading sedimentary rock, and while he denied hindering McLean's promotion at Virginia Tech, he said, "If the president of the college had asked me what I thought about Dewey McLean, I'd say he's a weak sister."

His primary antagonist in the piece was Robert Jastrow, a Dartmouth astronomer who delivered a harsh verdict against the asteroid hypothesis, while also alluding to Alvarez's participation in Hiroshima and the Oppenheimer hearing. Alvarez, in turn, criticized Jastrow's support of the Strategic Defense Initiative: "Jastrow, of course, has gotten into the defense of Star Wars, which for me personally indicates he's not a very good scientist."

On reading these remarks, the volcano advocate Charles Officer thought that Alvarez came off as "psychotic." In the article itself, Alvarez attributed his bluntness to a recent diagnosis for cancer: "This is my last hurrah, and I have to tell the truth." He had once expected to live as long as his father, but he turned out to have much less time, as a catastrophe of his own changed the situation overnight.

The previous August, he had lost his balance on Telegraph Avenue. A scan revealed a tumor the size of an egg at the base of his brain, surrounding the nerve that controlled his facial muscles. After it was removed, Alvarez awoke with the left side of his face drooping from Bell's palsy. Another procedure allowed his eyelid to close properly, but the effects were still visible, as he wrote in a letter to friends: "So you might as well get used to seeing me like that."

A few weeks later, he lost his appetite, and a tube inserted into his stomach produced only a slow leak of blood. He gathered his children for a final goodbye: "I thanked them for all they had done for us, wished them success in their lives, and felt confident that I would never see them again." The following morning, the tube was working well enough to save him from starvation. A biopsy of his esophagus revealed a malignant lesion, but he was tempted to simply accept the inevitable.

He was finally persuaded to undergo surgery on January 18, 1988, the day before the *Times* article was published. The surgeon shortened his esophagus by six inches, attaching his stomach to what was left. After additional operations and months in and out of the hospital, he wrote, "I am weaker than I have ever been in my adult life, and my main job is to get out of my bed."

Although he hoped to attend the annual Bohemian Club encampment, he was nearing the end. "The surgeries left him physically impaired and emotionally despondent," Trower wrote. "He presented his surgeon with a graph, the quality of life on the vertical axis and time on the horizontal, demonstrating that what remained for him to endure was not worth it." Previous periods of inactivity had always produced new ideas, but he no longer had it in him to fight.

He looked for reasons to be cheerful. At their observatory at Palomar Mountain, Gene Shoemaker and his wife, Carolyn, were searching for asteroids. One evening, they found an object over eight miles in diameter—around the size hypothesized for the asteroid in the K–T extinction—that they had the privilege of naming "Alvarez," honoring both father and son.

At times, Alvarez must have wondered if his cancer was due to radiation exposure, which had claimed von Neumann and Fermi at an early age. The possibility wasn't openly discussed, and his son Don later doubted it: "My guess has always been that with forty years between exposure and symptoms the odds of a generic origin for the condition were far higher than the odds of a historically interesting one." But no one would ever know for sure.

Don, a recent graduate of MIT, came home to spend time with his father, who liked to watch *Wheel of Fortune*. They noticed that if a contestant continued to guess after the solution was obvious, the wheel often landed on "Bankrupt," which implied that it might be rigged. At Don's suggestion, they taped the episodes, calculating the dynamics of the spin one frame at a time, but found that the camera usually cut away at the crucial moment to Vanna White.

As Don observed, this was how Alvarez connected with his sons, as well as with other important players in his life. Few ever experienced it at a deeper level than Muller, who visited him in the hospital. Seeing that Alvarez was concerned about losing his mental acuity, Muller gave him a toy that produced an optical effect—a moiré interference pattern—and said that he couldn't figure out how it worked. The next day, Alvarez informed him that he had found the answer. "That was my greatest gift to him," Muller recalled of their last meeting. "To let him know that he could still solve problems that I couldn't."

IN HIS MEMOIRS, Alvarez said of his religious beliefs, "To me the idea of a Supreme Being is attractive, but I'm sure that such a Being isn't the one described in any holy book. Since we learn about people by examining what they have done, I conclude that any Supreme Being must have been a great mathematician. The universe operates with precision according to mathematical laws of enormous complexity. I'm unable to identify its creator with the Jesus to whom my maternal grandparents, missionaries in China, devoted their lives."

On September 1, 1988, Alvarez died of complications from his operations for cancer. He was seventy-seven. After he was cremated, his remains were entrusted to Peter Trower. "A few days later," Trower wrote, "a small, heavy cardboard box in hand, I boarded a single-engine Cessna, instructed the pilot to fly out across the Golden Gate Bridge on to the Farallon Islands and there to head, as Luie had done many times piloting his own plane, to Monterey. I then scattered his ashes."

In his book *Wonderful Life*, which he was writing when he learned that Alvarez was gone, Stephen Jay Gould praised him as "a prince of science at the highest conventional grade." Unlike most scientists of his stature, he knew that he would be remembered for a breakthrough at the very end of his career. Six weeks after his death, at the second extinction conference in Snowbird, Utah, an attendee proposed that they honor Alvarez with two minutes of silence. Walt rose to speak. "My father would have been mortified. He'd much rather have a good fight in his memory."

EPILOGUE

So complex is reality, and so fragmentary and simplified is history, that an omniscient observer could write an indefinite, almost infinite, number of biographies. . . . A history of a man's dreams is not inconceivable; another, of the organs of his body; another, of the mistakes he made; another, of all the moments when he thought about the Pyramids; another, of his dealings with the night and with the dawn.

—Jorge Luis Borges

WALT NEVER GAVE UP HOPE THAT THE CRATER WOULD be found. When asked, he was always prepared to list the reasons that it might be impossible. In all likelihood, the asteroid or comet had fallen into the ocean, and even a land strike might have been erased long ago by sediment or ice. All the same, nothing would do more to silence his critics, and he never imagined that the answer, hiding in plain sight, could have been identified long before his father's death.

When the breakthrough came, it was due to Alan R. Hildebrand, a Canadian graduate student at the University of Arizona. Hildebrand was determined to discover the crater, based on the premise that an ocean impact would have produced a gigantic tsunami—over half a mile high—that left traces in its wake. These deposits, known as turbidites, would have been formed by the slurry of coastal sand torn up and carried overland by the monster wave.

Combining the known turbidite sites with the increasing thickness of the K–T layer to the south, he narrowed down the location to somewhere in the Gulf of Mexico or Caribbean. He scrutinized maps for a circular structure that might mark a crater's edge, but failed to find it. At a conference in Houston in 1990, however, he met a science reporter named Carlos Byars, who had written a newspaper article, almost a decade earlier, about an intriguing candidate in the Yucatán Peninsula.

When Hildebrand looked into it, he was astonished. In the fifties, Pemex, the state petroleum company of Mexico, had been exploring for oil. A gravity survey—designed to uncover variations in the density of the crust—found a feature like a horseshoe on the north coast of the Yucatán. Preliminary drilling extracted cores that resembled volcanic rock, which was unlikely to produce fossil fuels, so the location was forgotten for a quarter of a century.

In the late seventies, the search was renewed by Antonio Camargo and Glen Penfield, who were working as geophysicists for Pemex. Studying magnetic data taken by planes above the peninsula, Penfield noticed an unusual semicircle. When paired with the gravity readings, it formed a bull's-eye over a hundred miles in diameter. Camargo and

Relief map of the Yucatán Peninsula, Mexico. The semicircular trough of the Chicxulub crater is highlighted in the second image. Jet Propulsion Laboratory.

Penfield concluded that it wasn't a volcano at all, but an ancient impact crater, buried under thousands of feet of sediment.

Although Pemex treated the results as proprietary, the two scientists were allowed to give a talk at the Society of Exploration Geophysicists in 1981, mentioning the possible relevance to "the late Cretaceous extinctions." In an unfortunate quirk of timing, the disclosure coincided with the first extinction conference at Snowbird, which meant that the researchers most interested in the problem were occupied elsewhere when the answer was revealed.

Penfield and Camargo were outsiders whose work didn't appear in the usual places, so the hint went unnoticed. An abstract in the conference program was overlooked, along with Byars's article in the *Houston Chronicle* and a piece in the magazine *Sky & Telescope*. According to Penfield, he even wrote to Walt, but he never received a response. At the time, Walt was focusing on the oceanic crust, not the land, so he didn't follow up on the lead.

When Hildebrand learned about the site, he nicknamed it "Chicxulub"—pronounced "Cheek-shoe-lube"—after a village near its center. The word, derived from a Mayan root meaning "horned," was variously translated as "the red devil," "the devil's flea," or "the place of the cuckold." Hildebrand wanted to examine rocks from the crater, but it was buried too deeply to be easily accessible, and the cores extracted by Pemex had evidently been destroyed in a fire.

Walt was equally excited about the possibility, which he investigated with his limited resources. In a book from the thirties, he found a description of a sandstone bed at the K–T boundary in Mexico—the closest known location to Chicxulub. In 1991, a team led by Jan Smit and Sandro Montanari found an astounding three meters of sand there, along with felted deposits of petrified wood in what otherwise looked like ocean sediment. The tsunami, they realized, had uprooted entire forests.

By the end of the year, the case for Chicxulub grew even stronger. A more thorough search of a Pemex warehouse in Mexico miraculously uncovered two of the missing core samples. Radiometric dating indicated that the rock was precisely the right age; its chemical composition matched tektites at other locations; and it contained both shocked quartz and high levels of iridium.

For Walt—and many others—it was "the smoking gun." Within a few years, the Alvarez hypothesis seemed like conventional wisdom, and later accounts treated it as inevitable. As Walt told the science journalist Elizabeth Kolbert, "Those eleven years seemed long at the time, but looking back they seem very brief. Just think about it for a moment. Here you have a challenge to a uniformitarian viewpoint that basically every geologist and paleontologist had been trained in, as had their professors and their professors' professors, all the way back to Lyell. And what you saw was people looking at the evidence. And they gradually *did* come to change their minds."

Unlike Alvarez, who could move on to the next problem, paleontologists had to live with the consequences. Stephen Jay Gould correctly saw that they needed to examine the region around the impact layer more thoroughly, looking for the "dinosaur in a haystack" that could fill out the fossil record. As they painstakingly searched each level, rather than relying on sampling, the evidence for an abrupt extinction became increasingly convincing. "The Alvarez theory made this unusual approach necessary," Gould wrote. "The new idea forced us to observe in a different way."

Since devastating impacts had occurred in the past, it also raised the question of how to avoid them in the future. NASA gave one solution a trial run with the Double Asteroid Redirection Test, launching a spacecraft on an intercept course with Dimorphos, a moon of the asteroid Didymos. On September 26, 2022, it smashed into its target, successfully altering Dimorphos's velocity and orbital radius. It indicated that an asteroid strike could indeed be prevented, as long as it was foreseen in time.

A few lonely voices continued to advocate for the volcanism hypothesis. While the Deccan Traps in India clearly predated the K–T event, however, the best available estimates implied that most of the eruptions occurred after the Cretaceous. Walt, for his part, still suspected that the impact had played a role in producing the greatest lava flows. His argument was attractive but unproven, and its appeal to common sense offered good reasons—based on the listener's predisposition—either to favor it or to question it more closely.

The Nemesis hypothesis fared less well. Muller's proposed solar companion was never found, not even by infrared surveys that could detect cooler stars. Scientists saw whatever they wanted in the periodicity data, which was alternately confirmed and dismissed. Tellingly, though, the volume of iridium deposited worldwide was more consistent with an asteroid than a comet, and no one could suggest a credible mechanism for how such asteroids might strike in a regular cycle.

While Muller never gave up hope, the most electrifying hypothesis of his career, and perhaps of all time, appeared unlikely to be correct. Evidence of impacts at other extinction events was similarly inconclusive, as Elizabeth Kolbert noted: "If twenty-five years ago it seemed that all mass extinctions would ultimately be traced to the same cause, now the reverse seems true. As in Tolstoy, every extinction event appears to be unhappy—and fatally so—in its own way."

Gould felt that Alvarez would have been "disappointed and disturbed" to have found the explanation for the end of the dinosaur era but not for any others: "Luis, as we well know, did not have an enormous or abiding interest in natural history as a descriptive science." As a result, Alvarez might well have been dissatisfied by the achievement that ensured his lasting fame. "To learn that he had become godfather to the contingent explanation of a great event," Gould said, "and not to the formulation of a general theory of mass extinction, would have left him unamused."

ALVAREZ WOULD HAVE JUDGED his legacy against the achievement of one man. Lawrence, he wrote, deserved to be honored as "the inventor of the modern way of doing science." The most valuable products of the Radiation Laboratory had always been articles and students, and Alvarez's own impact consisted less of any one discovery than of the culture that he passed down to his protégés. Its success was measured in Nobel Prizes, which were produced in such profusion that a row of parking spaces was reserved for laureates at Berkeley.

A year after Alvarez's death, the Cosmic Background Explorer satellite went into orbit. It was a direct descendant of Muller and George Smoot's observations of the cosmic microwave background, which had benefited enormously from Alvarez's support. Freed from atmospheric interference, it found primordial variations, or "wrinkles," in radiation from the Big Bang, the faint seeds that enabled large structures—from galactic superclusters on down—to come into existence. Along with John C. Mather, Smoot was awarded the Nobel Prize in 2006 for what Stephen Hawking hailed as "the most important discovery of the century, if not of all time."

Five years later, the prize went to three other physicists, including Muller's former graduate student Saul Perlmutter. His work had arisen from Muller's proposal to use supernovas as standard candles, which Alvarez had advised implementing as an automated search. In the eighties, Perlmutter and Carl Pennypacker hoped to extend it to galaxies billions of light years away, allowing them to see how the universe's expansion was decelerating.

Muller, the group leader, felt that their plan left key problems unsolved, but he approved it anyway. "I had learned from Luis Alvarez that such daring was necessary—or you'd never tackle a great challenge." Alvarez had taught him to trust that any missing components would be invented in time, which Perlmutter confirmed: "One of the things [Muller] had learned from Luie was the sense that you have to support these ideas when they come up."

Perlmutter arrived at an approach that resulted, astonishingly, in the opposite of what they had expected. Instead of slowing down, the universe was expanding at an accelerating rate, driven by an unexplained "dark energy." It was one of the most startling discoveries in the history of astrophysics, and Perlmutter credited Alvarez's "cowboy spirit" for creating the culture that produced it: "As a physicist, you had the hunting license to look at any problem whatsoever."

Alvarez's tools had a productive afterlife of their own. Birger Schmitz, one of his last postdocs, used his iridium coincidence spectrometer to support the hypothesis that a distant asteroid breakup triggered an Ordovician ice age. Muon tomography—originally developed for the Pyramid of Chephren—was deployed to explore volcanoes and evaluate damage after the nuclear meltdown in Fukushima, Japan. Best of all was a collaboration between Egypt and France that began in 2015. Using Alvarez's methods, it detected three previously unknown voids in the Great Pyramid of Cheops, including one that was nearly a hundred feet long.

Although the chamber's purpose remained a mystery, it confirmed that Alvarez had been on the right track, even if he had chosen the wrong pyramid. Given the presence of other rooms and the consensus about construction methods, Cheops had offered the better odds of success, even at the time, while Chephren held out a higher reward at a greater risk. Given his personality, Alvarez's decision was unsurprising, but Jerry Anderson, the former head of the pyramid team, admitted, "I wish we had worked in the Great Pyramid."

Alvarez would have been less gratified by the ongoing debate over nuclear weapons. Throughout his life, he defended the destruction of Hiroshima and Nagasaki, especially as he learned more about the American plans for a ground invasion. He argued that any other choice would have been politically impossible, and when students pushed back on this point, he tested it in characteristic fashion—by consulting an original source at the library.

Without any prompting, he revisited the question while working on his autobiography with Richard Rhodes. "I didn't ask Luie to justify the bombing of Hiroshima and Nagasaki," Rhodes wrote, "I wasn't that presumptuous, but he'd been challenged on the question often enough that he brought it up himself, turned red in the face and slapped his hand on his desk and insisted the atomic bombings were necessary as a firebreak to shock the Japanese into surrender."

Alvarez advised Rhodes to read the last issue of *Life* published before Hiroshima. In an article titled "A Jap Burns," a full page was devoted to six graphic photos from Balikpapan, a seaport on Borneo. After an Australian soldier used a flamethrower on a Japanese encampment, a man emerged, skin on fire, and burned to death in full view of the camera. The text concluded darkly, "So long as the Jap refuses to come out of his holes and keeps killing, this is the only way."

If such horrific pictures could appear in the nation's most popular magazine, Alvarez said, it implied that the United States would have rejected anything short of unconditional surrender. In his memoirs, he imagined President Truman authorizing an invasion, leading to massive casualties, only for the public to discover that a weapon existed that could have ended the war much earlier: "How would Truman have explained his decision at his impeachment trial?"

Not surprisingly, Alvarez disagreed with Oppenheimer's statement that "the physicists have known sin," writing that using the bomb was "one of the most lifesaving decisions in the history of mankind." He also supported the arms race that followed, but was less hawkish than many of his contemporaries. Like his friend Arthur C. Clarke, he felt that the Strategic Defense Initiative, which he studied with the Jasons, was doomed to failure. While he was skeptical of disarmament talks, he thought that nuclear stockpiles should be reduced—although not eliminated.

In his autobiography, Alvarez wrote, "I can think of nothing more horrible than the sudden elimination of all nuclear weapons, which could, in my opinion, lead quickly to the invasion of Western Europe by the Soviets, and a death toll in the tens of millions." Alvarez stated flatly that "the present stability of the world rests primarily on the existence of nuclear weapons," and an earlier draft unequivocally credited "hydrogen bombs" with preventing a cataclysmic conflict: "I have frequently differed with Edward [Teller], but I am convinced that he, more than anyone who has ever lived, deserves the Nobel Peace Prize."

He claimed to "enthusiastically support" the doctrine of mutual assured destruction. Describing himself "as an acknowledged expert on the major catastrophes in the geological record," he wrote that deterrence had led to a reprieve from the cycle of world wars, allowing the

United States and Russia to move toward a position of trust. He predicted that the process would take another century, until the grandchildren of the "old soldiers" had all passed away. It was a strategy of postponement that needed time to work, "and time is what nuclear weapons have given us."

Alvarez died three years before the dissolution of the Soviet Union, which he would have seen as a vindication. The historical verdict is less clear. At the time of writing, there are nine nuclear powers, including Israel, which has never acknowledged its arsenal's existence, let alone any role in the Vela incident. On the threat of proliferation, Alvarez once said that terrorists could detonate a bomb of purified uranium "simply by dropping one half of the material onto the other half." After Russia invaded Ukraine in 2022, the Bulletin of the Atomic Scientists set its Doomsday Clock—a symbol of the probability of global disaster—to a minute and a half until midnight.

The tools of physics, Alvarez wrote, worked best within a narrowly defined range of experience: "When we physicists are confronted with a complicated problem, we usually set up a simplified model of the real system, and then proceed to solve the simpler problem." Over the course of his career, however, he constantly expanded his argument for what physics could do, encouraged by his success in the extinction debate. He once asked a colleague, "We know the basic physical laws of atoms and molecules, so what's holding up your understanding of the physics of biological processes? After all, cells are made of atoms and molecules."

In reality, life—like human society—resisted such straightforward laws. Alvarez correctly saw nuclear weapons as different from previous challenges, largely because of the destructive force of the hydrogen bomb: "Most people have no comprehension of what a change of a factor of ten million means, so it is not surprising that they use history as a guide to their thinking. In physics, we are accustomed to such large changes, and we know that the emergence of a 'factor of ten million' requires entirely new thinking 'from first principles'—not from history."

Alvarez used this argument to contend that the use of the bomb wasn't inevitable, but it actually pointed to a more difficult realization. As it happened, two Berkeley professors were examining these issues at exactly the same time, in a building just a short walk from the physics department. At the College of Environmental Design, Horst W. J. Rittel and Melvin M. Webber published a paper in 1973 titled "Dilemmas in a General Theory of Planning." They drew a distinction between the "tame" problems that scientists tended to address, with clear rules and objectives, and "wicked" problems that defied the tools of engineering.

One example of a wicked problem was what Alvarez called "severe climactic changes," which he identified as a looming crisis in the early seventies: "There is real danger that by burning coal, we will put so much extra carbon dioxide into the atmosphere that the 'greenhouse effect' will warm the earth to the point that the Antarctic icecap will melt, and all of the coastal cities will find themselves fifty or one hundred feet under water." He felt that the only solution lay in nuclear power, which would also reduce dependence on oil from the Middle East.

In practice, climate change—and the arms race—met all of Rittel and Webber's criteria for a wicked problem. Answers weren't true or false, but better and worse; there was no obvious stopping point; actions, once taken, couldn't be undone; and one bad outcome would negate all previous efforts. As a result, they wrote, "It becomes morally objectionable for the planner to treat a wicked problem as though it were a tame one, or to tame a wicked problem prematurely, or to refuse to recognize the inherent wickedness of social problems."

Alvarez's embrace of mutual assured destruction—with its immense political, social, and psychological costs—exemplified the limitations of seeing the postwar world through the lens of previous answers. It was less a lack of imagination than a perversion of the idea of a "solution" itself. Faced with the forces unleashed at Los Alamos, even geniuses could seem powerless. Creativity and intelligence were still necessary, but far from sufficient, and they were often used to enable the machine to continue along its predetermined course.

As Oppenheimer learned firsthand, there could be tragic consequences for scientists who pushed back. At times, Alvarez and his contemporaries seemed to be going through the motions, consoling themselves by endorsing an outcome that had already been decided. Rittel and Webber felt that the entire process was a charade: "The social professions were misled somewhere along the line into assuming they could be applied scientists—that they could solve problems in the ways scientists can solve their sorts of problems. The error has been a serious one."

IN THE UNCUT DRAFT of his autobiography, Alvarez acknowledged the temptation to look back fondly on a time when "a single experimenter did everything in his own laboratory room and published the results himself." Almost at once, however, he qualified this sense of nostalgia: "Experimental physics in the good old days was often inefficient, lonely, and frustrating, an ordeal that could mold the physicist's character into one not far from that of the horror movies' mad scientists."

At Berkeley, Alvarez recalled, Lawrence created a collaborative haven that "stimulated my imagination beyond all my expectations." It was a golden mean that he spent much of his career trying to recapture, conceiving of projects for small teams—the balloons, the pyramids—that evoked that lost sense of excitement. As experimental physics inexorably expanded in scale, this became harder to achieve, awakening him to the dangers of a different sort of isolation.

Ironically, it was a logical consequence of Lawrence's "big science," which had triumphed decisively in World War II. Afterward, Alvarez lamented that it was no longer possible for ideas to quickly secure government approval: "On the basis of my experience in wartime, I'm sure we waste far greater sums of money on bureaucratic review, delay, and red tape than we could possibly lose if a project occasionally failed or a dishonest outfit defrauded. In the process we squander the enthusiasm of the originators, which may be an even greater loss."

At the lab, Alvarez learned how to initiate ambitious programs with comparatively limited resources. His associate Stan Wojcicki—whose daughter Susan later converted her garage into office space for a tech startup called Google—recounted a piece of advice that may have said more than Alvarez intended: "He contrasted trying to become a good physicist with trying to become the president of General Motors: the latter was a political task requiring great care not to offend key people, while the most important ingredient in becoming a good physicist was curiosity."

This was only part of the truth. Alvarez felt that he succeeded "because of the interpersonal skills I have learned," and he was careful to emphasize the last word. He was well aware that experimentalists—with their need for funding—were more vulnerable to financial and political considerations than theorists. Following the examples of Lawrence and Oppenheimer, he tackled the problem systematically, applying the principles of capitalism to science.

One result was extreme specialization, accelerated by the technology that Alvarez introduced: "I brought the computers in." Young researchers, he saw, were adept at statistical analysis, but couldn't find their way around a machine shop. Reflecting on a truism that paper cost less than brass—meaning that it was best to work out ideas before building equipment—he noted that computers followed different economic rules: "[A student] ends up with a stack of computer printout that costs as much as many of the brass objects I have machined in my life. . . . I mention this only to show you how completely 'screwed up' our field of physics has become."

The computer upended the sense of scale in physics as dramatically as the hydrogen bomb had in warfare, but the process of expansion had begun much earlier. Lawrence had made sure that the Bevatron was big enough to discover the antiproton, and more massive particles demanded even greater energies. Alvarez lived long enough to see the construction of the Tevatron at Fermilab, the most gigantic accelerator of its time, which still wasn't enough for the final step. Producing the Higgs boson, the last unobserved particle predicted by theorists, had to wait for the Large Hadron Collider near Geneva, which was finished for $9 billion in 2008.

As such gargantuan projects limited the scope for personal achievement, Alvarez was concerned that the best students would prefer disciplines "where important work can still be done in teams smaller than ten," which was no longer true of his field. "Most of us do physics because it's fun and because we gain a certain respect in the eyes of those who know what we've done. Both of those rewards seem to me to be missing in the huge collaborations that now infest the world of particle physics." Observing that recent papers had dozens of authors, he doubted that he would have "derived any satisfaction" from appearing halfway down a list of contributors.

Another consequence was the dominance of theory, where it was relatively easier for an individual to reach a prominent position. Theorists on laboratory scheduling committees authorized searches for particles that their models predicted, while vetoing the "crazy experiments" that Alvarez thought should take up ten percent of an experimenter's time. He criticized peer review of proposals as "the greatest disaster to be visited upon the scientific community in this century."

Taking pride in his reputation for achieving results "in more fields than almost any other scientist," he believed that the existing system discouraged researchers from changing direction or crossing professional boundaries, which had produced many of his own breakthroughs. Physicists spent their lives repeating their graduate work, knowing that they would lose funding if they moved to another department: "That's enough to keep almost anyone from leaving his own sandbox."

To encourage innovation, it was necessary to consider the experimenter's track record, not just the experiment. When asked how a student could build a reputation, Alvarez said, "He would do it the same way I did; I apprenticed myself to Ernest Lawrence, a man with such a good track record that any proposal he made was automatically funded without peer review. After a few years, if he had any talent, my young scientist would have his own track record and assurance of support when his peers reviewed him—not his proposal."

Apprenticeship was undoubtedly crucial for experimental physicists, who needed the unwritten knowledge—what Richard Rhodes called the "tricks of the trade"—that Alvarez had acquired "only through informal conversations with men who have spent years with the subject." Yet his own story indicated that this process could be less than straightforward. "I believe I'm widely reputed to be approachable and friendly," he wrote in his memoirs, which was ludicrously off the mark. He had a very different reputation in Berkeley, largely because he had trouble expressing criticism that didn't feel like a personal attack.

As Alvarez knew well, his ability to attract talented collaborators was a powerful asset: "You don't want to be playing with the scrubs, you want to be in the first team, you want to be working with the people that you admire most." Yet many associates remained loyal to him despite his personality, not because of it. Even if his difficult side was inseparable from his accomplishments, it called for sacrifices that not every potential scientist was capable of making.

In particular, the system worked to the disadvantage of women, people of color, and anyone without the safety net of privilege that allowed Alvarez to flourish and take risks. His first job with Lawrence was due primarily to his father's role in securing grants for the lab. As his wives took care of the household labor, he was free to devote himself to research, as well as the luxury of sustained contemplation, and he could wait patiently for opportunities that were unimaginable to others.

Mentorship itself required a degree of sensitivity that had to be learned, and Alvarez never gave it the same attention that he lavished on so many other skills. He might have argued that this uncompromising treatment was the price of science, but it was harder to pin down what—or who—was lost along the way. Although he occasionally encouraged protégés who didn't look like him, it was easier to favor students like Muller, who consciously modeled himself on Alvarez.

Alvarez felt that his success as a leader came largely from his ability to guide others "into dangerous but rewarding new territory." He held himself to the same high standard, and he was willing to put his reputation on the line. "I've long felt a grudging admiration for Hernando Cortés. He was a despicable person, but by burning his ships to the waterline when he landed his small band of warriors in Mexico he showed them he was completely committed to its conquest."

Even for those who lived up to his expectations, however, the rewards were far from guaranteed. His finest protégé, Rich Muller, never won the Nobel Prize, which went instead to his mentor, rival, and former student. Part of the problem, Muller conceded, was that he had trouble sticking with ideas, and chance played a role as well: "Learning how to pick the right projects is a little bit like learning to pick the right parents. It helps to be in the right place at the right time."

In a famous essay on Tolstoy, the critic Isaiah Berlin drew a distinction between two types of geniuses, based on a fragment from the Greek poet Archilochus: "The fox knows many things, but the hedgehog knows one big thing." By Berlin's definition, Alvarez—and Muller—were among the foxes who "lead lives, perform acts and entertain ideas that are centrifugal rather than centripetal."

Rather than rewarding the foxes, the future seemed to rest with hedgehogs, who could single-mindedly pursue one idea for years, as embodied by another product of the Berkeley lab. Alvarez presciently wrote in the eighties, "Smoot is content to spend half his life on this one project, and he will do an excellent job, but that is not Muller's style." As complex problems continue to call for the combined effort of many specialists, the generalist might feel like a romantic anachronism.

In reality, both types are necessary, and they arise from the same place. "Creativity in science only comes *after* one has worked hard at the dull routine of learning a lot of basic skills in mathematics and physics," Alvarez said, "and in the fundamentals of one's particular science of choice." The danger, as he recognized, is that the acquisition of information can become an end in itself. Knowledge—like inventiveness—is meaningful only when tested by important problems.

Another indispensable factor is collaboration. While the discovery of the cause of the K–T extinction is often seen as a case study in the potential rewards when an outsider enters a new field, it succeeded only because Alvarez's son was waiting for him there. As Alvarez noted, it demanded "Walt's geological expertise, Frank's and Helen's nuclear and chemical competence, and my background in physics and astronomy." Otherwise, it might not have occurred at all: "Discoveries are often made simultaneously by two or more groups because the time is right for them. I don't believe this was the case with the impact hypothesis."

Alvarez's example only underlines the challenges of the modern age, even for a man of his gifts. Wicked problems can't be addressed solely by individuals like him, but they can never be solved without them. And although Alvarez presumably would have agreed, he was less than optimistic about his place in the world that he and his collaborators had made. As he wrote toward the end of his life, "There is no way that a person with my personal qualities could go into either nuclear physics or particle physics at the present time."

ACKNOWLEDGMENTS

I OWE THE MOST to Walter, Jean, Don, and Helen Alvarez. Thanks as well to Catherine Asaro, Lina Galtieri, Andrew Harth, Oscar Harth, Bill Higgins, Paul Hoch, Dan Hooper, Stephan Hruszkewycz, Alan Jackson, Jack Lloyd, John C. Mather, Richard A. Muller, Carl Pennypacker, Saul Perlmutter, Richard Rhodes, Raphael Rosen, Robert J. Sawyer, Birger Schmitz, Jan Smit, George Smoot, Anthony Spadafora, and Cliff Stoll. A special debt of gratitude is due to my research assistants Kazuyuki Murphey, Gavin Lee, and Brendan Williams-Childs.

Indispensable support was provided by Iris Donovan, Fedora Gertzman, and the entire staff of the Bancroft Library at UC Berkeley; Allison Rein at the American Institute of Physics; Joe DiLullo at the American Philosophical Society; Christine Colburn, Kathleen Feeney, Pete Segall, and Alexa Tulk at the University of Chicago; Shelby Schellenger and Kathy Lafferty at the University of Kansas; John German, Alice Muller, Carol J. Burns, and Matt Nerzig at Lawrence Berkeley National Laboratory; Todd Nichols, McKenzie Vaupel, and the staff of the Los Alamos Historical Society; Brye Anne Steeves at Los Alamos National Laboratory; Joanie Gearin at the National Archives at Boston; Caitlin McShea at the Santa Fe Institute; and George R. R. Martin and Sid Khalsa in Santa Fe.

This book was funded in part by a grant from the Alfred P. Sloan Foundation, where Doron Weber and Shriya Bhindwale have been unfailingly helpful. Matt Weiland is the best editor I've ever had, and I remain endlessly impressed by Yumiko Gonzalez Rios, Huneeya Siddiqui, Sarah Johnson, Brian Mulligan, and the team at W. W. Norton. My agent, David Halpern, has been a tireless champion and advisor, and I feel lucky to have worked with Kathy Robbins and Janet Oshiro at the Robbins Office and Jon Cassir at CAA. Thanks and love to my parents in Castro Valley; to my brother and his family in New York; to the Wongs; to my friends in Oak Park and elsewhere; and above all to Wailin and Beatrix.

NOTES

Box and carton numbers refer to the Luis W. Alvarez papers (BANC MSS 84/82 cz) at the Bancroft Library at the University of California, Berkeley. Uncited quotations from Alvarez appear in the published text of *Alvarez: Adventures of a Physicist.* Unsourced quotes from living persons are from author correspondence or interviews. Endnotes with all citations can be downloaded at www.nevalalee.com.

ABBREVIATIONS

AIP1 / AIP2 — American Institute of Physics interview with LWA, February 14–15, 1967, Sessions 1 and 2, https://www.aip.org/history-programs/niels-bohr-library/oral-histories/4483-1.

AR — "Alvarez, Luis W.," Anne Roe papers (Mss.B.R621), American Philosophical Society, Philadelphia, PA.

BL1 — Bancroft Library draft of LWA autobiography, C3.

BL2 — Bancroft Library draft of LWA autobiography, B46.

DA — LWA, *Discovering Alvarez.*

HC — Herbert Childs interview with LWA, "Materials Assembled for a Biography of Ernest O. Lawrence" (CU-369), C1-F5, Bancroft Library.

JRO — In the Matter of J. Robert Oppenheimer (hearing transcript), Washington, DC: Atomic Energy Commission, 1954.

LWA — Luis W. Alvarez.

TV — "Trower version" of LWA autobiography, Richard Rhodes papers (RH MS 1134), B34-F48, University of Kansas, Lawrence, KS.

EPIGRAPH

ix "Gossip is the backbone": LWA to Herbert York, April 23, 1985, B47.

PROLOGUE

1 **"Successful physics"**: BL2, 107.
3 **"A Matter"**: *Life*, November 25, 1966, 38–53.
3 **"I got very little sleep"**: LWA, "A Physicist Examines the Kennedy Assassination Film," *DA*, 211.
3 **"The head snap"**: Zapruder, *Twenty-Six Seconds*, 202.
5 **"tepid"**: Charles G. Wohl, "Scientist as Detective," *American Journal of Physics* 75, no. 11 (November 2007): 973.
6 **"Alvarez was a white"**: Rubén Martínez, "Adventures of Luis Alvarez," vii.
6 **"prize wild-idea man"**: "Bright Spectrum," *Time*, November 18, 1957.
6 **"the most creative"**: Seaborg, *Adventures in the Atomic Age*, 158.
6 **"Luie Alvarez, I think"**: "Alvarez Symposium—Rich Muller," YouTube video posted by Philip Dauber December 28, 2011, https://www.youtube.com/watch?v=_5dDUjoZMe4.
6 **"But then I think"**: Feynman, *What Do You Care What Other People Think?*, 146.
7 **"The only correct"**: LWA, textbook draft, C3, I-2.
7 **"In the fields"**: Louis Pasteur, University of Lille lecture, December 7, 1854.
7 **"killer instinct"**: Richard A. Muller, "An Adventure in Science," *New York Times*, March 24, 1985.
7 **"a sort of genius"**: Arthur Rosenfeld in Chuck Levy, "Alvarez the Nobelist," *California Monthly*, January/February 1969.
8 **"If someone thinks"**: Anne Roe interview (ca. 1952), AR, 6.
8 **"That's what they"**: Unidentified source in Roe, *Making of a Scientist*, 205.
8 **"Although Luie"**: Robert D. Watt, "Life with Luie," *DA*, 108.
8 **"Luie had taught"**: Muller, *Nemesis*, 33.
8 **"a knack"**: Muller, 27.
9 **"You have a greater"**: BL2, 429.
9 **"Oh, no"**: Rosa Segrè in Segrè, *Mind Always in Motion*, 297.
9 **"Luis Alvarez—wasn't he"**: Greenstein, *Portraits of Discovery*, 103.
10 **"He imitated"**: Davis, *Lawrence and Oppenheimer*, 86.
10 **"a tense"**: Davis, 301.
10 **"Fantastic ego"**: Davis, 253.
10 **"hard-headed"**: Davis, 313.
10 **"a healthy bit"**: "Alvarez Symposium—Saul Perlmutter," YouTube video posted by Philip Dauber December 28, 2011, https://www.youtube.com/watch?v=XG7tIs5HYiE.
11 **"In the whole"**: Thompson, *Last Second in Dallas*, 115.
11 **"consorting"**: BL2, 837.

PART I

13 **"We cannot"**: Ralph Waldo Emerson, "Napoleon," *Representative Men* (Boston: Phillips, Sampson, 1850).

CHAPTER 1

15 **"termites"**: BL2, 10.
16 **"I wonder if he"**: BL2, 10.
16 **"You nearly died"**: Alvarez, *Incurable Physician*, 54.
17 **"The thin bones"**: Alvarez, 56.
17 **"a Swiss Jew"**: Alvarez, 52.
17 **"If her name"**: "Don't Fret, Columnist Advises," *Oakland Tribune*, October 21, 1954.
17 **"going crazy"**: Alvarez, *Incurable Physician*, 73.
18 **"what was wrong"**: BL2, 56.
18 **"He had always"**: Anne Roe notes, March 1952, AR, 1.
18 **"A child who"**: Alvarez, *Incurable Physician*, 26.
20 **"losing my precious ass"**: LWA to Martin Kamen, July 9, 1985, B47.
20 **"In the eyes"**: Slosson, *Creative Chemistry*, 14.
22 **"a chemist's war"**: BL2, 51.
22 **"I learned"**: BL2, 27.
24 **"trouble"**: BL2, 53.
24 **"Hoag rewarded"**: BL2, 55.
24 **"Why don't you"**: AIP1.
24 **"a brash kid"**: BL2, 62.
26 **"This misapprehension"**: BL2, 70.
26 **"wizards or sorcerers"**: AIP1.
26 **"the mad Spaniard"**: Rubén Martínez, "Adventures of Luis Alvarez," 258–59.
26 **"the Spanish Swede"**: Rubén Martínez, 259.
27 **"unspoken consent"**: BL2, 56.
27 **"It did not occur"**: BL2, 56.
27 **"My scientific friends"**: BL2, 75.
30 **"the most brilliant"**: Walter C. Alvarez, autobiography pages, May 7, 1962, Personal 1952–1964, C1.
30 **"clamp cartel"**: BL2, 101.
30 **"There wasn't much"**: BL2, 108.
30 **"Do you know"**: BL2, 110.
33 **"I'm Gladys Archibald's brother"**: HC, 2.
35 **"like a Tennessee Williams play"**: AIP1.
35 **"the enthusiasm"**: Heilbron and Seidel, *Lawrence and His Laboratory*, 242.
35 **"The methods"**: Katharine Graham, "From Earliest Childhood," *Washington Post*, January 25, 1997.
36 **"a splendid idea"**: Lawrence to LWA, July 8, 1935, Alvarez family papers.
36 **"The reason"**: LWA, "Adventures in Nuclear Physics" (Faculty Research Lecture), March 1962, 12.
36 **"Lawrence, who was"**: "Professor Ryokichi Sagane," C1, 417.

CHAPTER 2

38 **"All he expected"**: AIP1.
38 **"I just felt"**: AIP1.
38 **"a country boy"**: LWA, *Alfred Lee Loomis*, 325.
38 **"From the discovery"**: LWA, textbook draft, C3, VI-1.
39 **"Simple calculations"**: Lawrence, "The Invention of the Cyclotron" (Nobel lecture), December 11, 1951.
40 **"to bombard"**: Childs, *American Genius*, 139.
44 **"into the room"**: BL2, 194.
44 **"I'll turn off"**: Childs, *American Genius*, 265.
44 **"would kneel"**: BL2, 236.
44 **"not very interested"**: LWA, *Ernest Orlando Lawrence*, 265.
45 **"I was two"**: Childs, *American Genius*, 262.
46 **"This was undoubtedly"**: BL2, 260.
46 **"a jack of all trades"**: AIP1.
47 **"coming of age"**: BL1, 335.
47 **"Most of the graduate"**: AIP1.
50 **"In addition"**: BL2, 250.
50 **"borrowed"**: BL2, 267.
50 **"We sat there"**: AIP1.
51 **"Ernest believed"**: BL2, 266.
51 **"her old Chinese nurse"**: Alvarez, *Incurable Physician*, 94.
51 **"I think I am behaving"**: Alvarez, 141.
51 **"I'm sitting"**: AIP1.
54 **"exciting physics"**: Childs, *American Genius*, 282.
54 **"his scientific sons"**: BL2, 311.
56 **"a pretty grim time"**: BL2, 300.
56 **"We each thought"**: LWA to Martin Packard, November 15, 1986, B39.
56 **"best pupil"**: LWA to Martin Packard, November 15, 1986.
59 **"A physicist"**: LWA, "Adventures in Nuclear Physics" (Faculty Research Lecture), March 1962, 21.
60 **"very clever"**: Segrè, *Mind Always in Motion*, 135.
60 **"[In 1937, Alvarez]"**: Raymond T. Birge, *History of the Physics Department*, vol. 4, XIV-11, HathiTrust, accessed June 2024, https://babel.hathitrust.org/cgi/pt?id=ucl.c047196332.

CHAPTER 3

62 **"I'm awfully"**: Bird and Sherwin, *American Prometheus*, 95.
64 **"not much of the Sanskrit"**: BL2, 315.
64 **"I'd never thought"**: HC, 36.
64 **"That's impossible"**: AIP2.
65 **"fishy"**: Anne Roe interview (ca. 1952), AR, 7.

65 **"[Abelson] was so"**: AIP2.
66 **"I don't remember"**: BL2, 322–23.
67 **"the most heavily"**: LWA to Louis Turner, August 19, 1940, B34.
70 **"I correctly assumed"**: BL2, 394.
70 **"that this great project"**: Hiltzik, *Big Science*, 225.
71 **"Radar won"**: Conant, *Tuxedo Park*, 284.
71 **"I'm very serious"**: Bird and Sherwin, *American Prometheus*, 212.
71 **MIT Radiation Lab:** See also LWA notebooks #11, #107, #297, M.I.T. Rad Lab, RG 227, National Archives (Waltham, MA).
73 **"We've done it, boys"**: Buderi, *Invention That Changed the World*, 104.
73 **"I had my words"**: Buderi, 104.
74 **"I don't know"**: LWA to Robert Shankland, December 18, 1980, B27.
76 **"I just predicted"**: AIP2.
76 **"leaky pipe"**: Buderi, *Invention That Changed the World*, 136.
76 **"Eagle was a good"**: *Five Years at the Radiation Laboratory*, 22.
76 **"Alvarez's folly"**: "Longhairs and Short Waves," *Fortune*, November 1945, 162–65.
77 **"The whole tempo"**: Rhodes, *Making of the Atomic Bomb*, 400.
78 **"[they] were not"**: Herken, *Brotherhood of the Bomb*, chap. 4, note 57, accessed June 2024, https://brotherhoodofthebomb.com/bhbsource/endnotes.html.
78 **"his oversized ego"**: Herken, 73.
78 **"cold fish"**: BL2, 528.
78 **"Gerry later extracted"**: BL2, 528.
79 **"Let's do the numbers"**: Richard A. Muller interview, April 26, 2023.
79 **"The only times"**: BL2, 733.
79 **"I never had"**: AIP2.
80 **"You are two miles"**: Solberg, *Conquest of the Skies*, 258. See also Chester Porterfield, "The Story of GCA," *Air Trails and Science Frontiers*, June 1947.
81 **"What a shock"**: BL2, 420.
81 **"for leading him astray"**: Handwritten notes on LWA interview, May 27, 1943, Coordination Committee Series 1, Box 49a/b, M.I.T. Rad Lab, RG 227, National Archives (Waltham, MA).
82 **"other father"**: BL2, 370.
82 **"human servos"**: Lawrence Johnston, "The War Years," *DA*, 60.
82 **"was in for"**: "Oral History: Kathryn and Charles Fowler," Engineering and Technology History Wiki, last edited February 24, 2023, https://ethw.org/Oral-History:Kathryn_and_Charles_Fowler.
82 **"Can you prove"**: Charles A. Fowler, "Rad Lab, Luie Alvarez, and the Development of the GCA Radar Landing System," *IEEE A&E Systems*, May 2008.
83 **"The bumpers"**: Fowler, "Rad Lab, Luie Alvarez."
84 **"I am assuming"**: LWA to Lawrence, April 16, 1943, Alvarez family papers.
84 **"What you and I"**: Lawrence to LWA, May 7, 1943, Alvarez family papers.
85 **"I never did"**: Hope, *I Never Left Home*, 25.
85 **"The engine's"**: LWA to "Gang" ("Round Robin" letter), December 26, 1943, Alvarez family papers.

86 **"The target"**: BL2, 470.
87 **"It would not have been possible"**: Clarke, *Astounding Days*, 214.
88 **"Dr. Wendt"**: Clarke, *Glide Path*, 106.

PART II

89 **"When you see"**: JRO, 81.

CHAPTER 4

91 **"a great shot"**: BL2, 479.
91 **"I have to say Einstein"**: Walter Alvarez interview, April 23, 2023.
93 **"I assume"**: AIP2.
94 **"Give me another"**: Libby, *Uranium People*, 144.
94 **"was high"**: Libby, 144.
94 **"certainly the finest"**: Libby, LWA blurb.
94 **"did several great"**: LWA to Fermi Award panel, October 14, 1981, B27.
94 **"scared a lot"**: Richelson, *Spying on the Bomb*, 41.
95 **"It really made"**: BL1, 872.
96 **"There is no democracy"**: Greenberg, *Politics of Pure Science*, 43.
97 **"Shall be in Chicago"**: Telegram from Oppenheimer to LWA, March 25, 1944, Alvarez family papers.
98 **"personal belongings"**: BL1, 887.
98 **"If you ever saw"**: HC, 22.
100 **"Fermi, like Groves"**: Hoddeson et al., *Critical Assembly*, 240.
101 **"This problem"**: BL2, 505.
101 **"a rainstorm"**: LWA in Hargittai and Hargittai, *Candid Science V*, 205.
102 **"He felt Dr. Oppenheimer"**: JRO, 786.
102 **"We've got to try"**: Hoddeson et al., *Critical Assembly*, 172.
103 **"We'll proceed"**: Coster-Mullen, *Atom Bombs*, 73.
105 **"I can go"**: "Early Day Experiences with Radiation Hazards," B21, 13.
105 **"by today's standards"**: Hoddeson et al., *Critical Assembly*, 150.
105 **"Now we have"**: Rhodes, *Dark Sun*, 155.
106 **"Luis Alvarez used"**: Albright and Kunstel, *Bombshell*, 82.
106 **"was allegedly"**: Jerome J. Maxwell, SAC (Albuquerque) FBI report, March 16, 1950, David and Ruth Greenglass FBI file.
106 **"secret inside stuff"**: Davis, *Lawrence and Oppenheimer*, 176.
107 **"Geraldine, who was normally"**: BL2, 529.
108 **"As a matter"**: BL1, 913.
110 **"Had we been looking"**: BL2, 541.
111 **"This fire ball"**: "Eyewitness Accounts of the Explosion at Trinity on July 16, 1945," AtomicArchive.com, accessed June 2024, https://www.atomicarchive.com/resources/documents/trinity/alvarez.html.

112 **"solvents"**: TV, 1.
114 **"We couldn't expect"**: BL2, 547.
114 **"get the hell out"**: BL2, 551.
115 **"Paul Tibbets and his commanders"**: BL2, 551.
115 **"each costing"**: TV, 4.
115 **"Great, just great"**: Thomas and Witts, *Enola Gay*, 231.
116 **"the opening"**: LWA, "Letter to Walter Alvarez," *DA*, 70.
118 **"It was a beautiful"**: Knebel and Bailey, *No High Ground*, 206.
118 **"We hope that"**: "Lawrence Johnston," *Spokesman-Review* (Spokane, WA), January 27, 1985.
118 **"This is the first"**: LWA, "Letter to Walter Alvarez," *DA*, 69.
119 **"more power than"**: "Truman Reports, 'It Is an Atomic Bomb,'" *Air & Space Forces*, September 1, 2006.
119 **"Headquarters, Atomic Bomb Command"**: "This Rain of Atomic Bombs Will Increase Manyfold in Fury," Letters of Note, December 14, 2009.
121 **"I have never been"**: BL2, 566.
121 **"a visual drop"**: BL2, 570.
122 **"almost neurotic"**: Knebel and Bailey, *No High Ground*, 250.
122 **"We can propose"**: Bird and Sherwin, *American Prometheus*, 299.

CHAPTER 5

125 **"it would be no less"**: Childs, *American Genius*, 371.
125 **"an essentially living art"**: "A Report on the International Control of Atomic Energy," Department of State, March 16, 1946, 6.
126 **"We both knew"**: BL2, 733.
126 **"We ran it"**: Hiltzik, *Big Science*, 311.
126 **"All of us"**: "History of Proton Linear Accelerators," B48, 1.
126 **"some easy way"**: LWA to Ernest Lawrence, July 7, 1945, Alvarez family papers.
128 **"perfectly reasonable"**: "History of Proton Linear Accelerators," B48, 9.
129 **"In recent weeks"**: BL2, 631.
130 **"running out"**: Herken, *Brotherhood of the Bomb*, 176.
130 **"which had the shortest"**: BL2, 243.
130 **"the world beater"**: Hargittai and Hargittai, *Candid Science V*, 209.
131 **"Now we have"**: W. K. H. Panofsky, "Building the Proton Linear Accelerator," *DA*, 75.
131 **"rather controversial"**: Panofsky, "Building the Proton Linear Accelerator," 76.
131 **"The linac was"**: LWA to George Rogosa, December 12, 1958, B4.
131 **"Science is like baseball"**: Waltz, *What Makes a Scientist?*, 133–34.
131 **"divided his talent"**: Davis, *Lawrence and Oppenheimer*, 253.
131 **"This served"**: Buderi, *Invention That Changed the World*, 415.
132 **"any party"**: Hiltzik, *Big Science*, 332.
132 **"Important faculty"**: BL2, 696.
133 **"Then you can no longer"**: Segrè, *Mind Always in Motion*, 235.

133 **"Ernest grunted"**: Hiltzik, *Big Science*, 334.
133 **"security person"**: Cole, *Something Incredibly Wonderful Happens*, 86.
133 **"lectured by Alvarez"**: Hiltzik, 335.
133 **"good chemistry"**: BL1, "Hardball Physics," 4.
133 **"Serber wrote"**: Hargittai and Hargittai, *Candid Science VI*, 739.
133 **"For theory"**: Segrè, *Mind Always in Motion*, 235.
134 **"Now that we have"**: Schwartz, *Last Man Who Knew Everything*, 192.
134 **"the program essentially"**: JRO, 775.
134 **"The only thing"**: JRO, 774. LWA diary quotes in this section are from JRO, 774–84.
135 **"I don't need"**: "The Decision to Go Ahead with the Hydrogen Bomb," Web of Stories, accessed June 2024, https://www.webofstories.com/play/edward.teller/98.
136 **"They declared"**: Rhodes, *Dark Sun*, 385.
136 **"bloodthirsty"**: Lilienthal, *Journals*, 582.
136 **"Ernest Lawrence and Luis Alvarez"**: Lilienthal, 577.
136 **"a quantum jump"**: Herken, *Brotherhood of the Bomb*, 203.
136 **"I generally find"**: JRO, 461.
137 **"As I remember"**: JRO, 784.
137 **"deputized"**: JRO, 784.
137 **"he thought I would"**: Rhodes, *Dark Sun*, 394.
137 **"If there were a better"**: Serber, *Peace and War*, 169.
138 **"I watched"**: JRO, 785.
138 **"Yes or no?"**: Davis, *Lawrence and Oppenheimer*, 313.
138 **"If we built"**: JRO, 785.
138 **"what other people"**: JRO, 247.
138 **"What does worry"**: Bird and Sherwin, *American Prometheus*, 419.
138 **"a weapon of genocide"**: Rhodes, *Dark Sun*, 401.
139 **"I think the reason"**: JRO, 787.
139 **"This was the first"**: JRO, 787.
139 **"frankly stared"**: Childs, *American Genius*, 422.
140 **"negative but not damning"**: Davis, *Lawrence and Oppenheimer*, 270.
140 **"Forget obstacles"**: Childs, *American Genius*, 386.
140 **"spectacular 'stalagmites' "**: Panofsky, *Physics, Politics, and Peace*, 41.
141 **"Anyone who now"**: Herken, *Brotherhood of the Bomb*, 233.
141 **"two years I wasted"**: TV, 109.
141 **"Today is a red-letter"**: Panofsky in Hargittai and Hargittai, *Candid Science VI*, 609.
141 **"Nobody ever does"**: Hargittai and Hargittai, 623.
141 **"high-handed position"**: Herken, *Brotherhood of the Bomb*, 235.
141 **"He remained silent"**: Teller, *Memoirs*, 306.
142 **"Ernest felt that Robert"**: HC, 34.
142 **"We are having"**: JRO, 787.
142 **"It will die"**: JRO, 788.
142 **"If it's important"**: HC, 27.
142 **"forced to the conclusion"**: Stern, *Oppenheimer Case*, 172.
143 **"Luis, how could you"**: JRO, 788.

143 **"It has caused me"**: JRO, 789.
143 **"Oppenheimer was, in effect"**: Stern, *Oppenheimer Case*, 172.
143 **"agitated"**: Herken, *Brotherhood of the Bomb*, 249.
143 **"supermen"**: Jungk, *Brighter Than a Thousand Suns*, 282.
144 **"Edward and I"**: BL1, 1161.
144 **"Ten megatons"**: BL1, 1161.
144 **"It's a boy"**: Teller, *Memoirs*, 352.
144 **"I nodded"**: JRO, 802.
144 **"multiplied the effective"**: BL1, 878.
145 **unidentified "paper"**: McMillan, *Ruin of J. Robert Oppenheimer*, 171.
145 **"More probably"**: Bird and Sherwin, *American Prometheus*, 478.
145 **"To foreclose"**: Bird and Sherwin, 491.
145 **"It was further reported"**: JRO, 6.
146 **"on the black list"**: Herken, *Brotherhood of the Bomb*, 283.
146 **"Alvarez said"**: Herken, 283.
146 **"Oppenheimer was a Russian agent"**: Herken, 283.
146 **"furious to explain"**: Rhodes, *Dark Sun*, 552.
146 **"Both men pointed"**: Teller, *Memoirs*, 371.
146 **"one of the leaders"**: Davis, *Lawrence and Oppenheimer*, 314.
147 **"a person of that kind"**: Rhodes, *Dark Sun*, 552.
147 **"Teller was most"**: Davis, *Lawrence and Oppenheimer*, 313.
147 **"I had never heard"**: BL2, 683.
147 **"my commitment"**: Herken, *Brotherhood of the Bomb*, chap. 17, note 88, accessed June 2024, https://brotherhoodofthebomb.com/bhbsource/endnotes.html.
147 **"let me have it"**: BL2, 684.
148 **"relating to the container"**: FBI memo, March 17, 1950, David and Ruth Greenglass FBI file.
148 **"somewhat hard-headed"**: FBI memo.
148 **"a little dubious"**: FBI memo.
149 **"intended to harm"**: SAC (San Francisco), June 16, 1948, LWA FBI file.
149 **"annoyed and embarrassed"**: Alexander E. Campbell, August 20, 1948, LWA FBI file.
149 **"quite the big shot"**: SAC (Los Angeles), April 16, 1951, LWA FBI file.
149 **"bragging campaign"**: SAC (Los Angeles).
149 **"a certain lack"**: Charles F. Brusch, September 25, 1951, LWA FBI file.
149 **"a third party"**: Bird and Sherwin, *American Prometheus*, 198.
149 **"treason"**: JRO, 130.
149 **"I have known"**: JRO, 146.
149 **"cock and bull story"**: JRO, 153.
150 **"It seems entirely"**: Bird and Sherwin, *American Prometheus*, 510.
150 **"possibly phonetic"**: Roberts, *Brother*, 198.
150 **"Schmell"**: SAC (Albuquerque), March 3, 1950, David and Ruth Greenglass FBI file.
150 **"the best suspect"**: FBI memo, March 17, 1950, David and Ruth Greenglass FBI file.

150 **"kidnap certain"**: Charles F. Brusch, March 23, 1950, LWA FBI file.
151 **"that among"**: FBI memo, March 17, 1950, LWA FBI file.
152 **"I also knew"**: BL2, 685.
152 **"people who were afterwards"**: Bird and Sherwin, *American Prometheus*, 533.
152 **"He certainly raised"**: Oppenheimer hearing quotations in this section are from JRO, 771–805.
155 **"loyal citizen"**: "Findings and Recommendations of the Personnel Security Board in the Matter of Dr. J. Robert Oppenheimer," Avalon Project, Lillian Goldman Law Library, Yale Law School, 2008, https://avalon.law.yale.edu/20th_century/opp01.asp.
155 **"the vital interests"**: JRO, 710.
155 **"rough treatment"**: Teller, *Memoirs*, 399.
155 **"notable exceptions"**: "Scientists—Politics," *El Paso (TX) Times*, June 22, 1954.
156 **"Oppie and I"**: HC, 41.

CHAPTER 6

157 **"midlife crisis"**: BL1, 1125.
157 **"modern particle physics"**: LWA, "Recent Developments in Particle Physics," *DA*, 110.
157 **"Who ordered that?"**: Panofsky, *Physics, Politics, and Peace*, 111.
158 **"[Alvarez] was the only"**: Davis, *Lawrence and Oppenheimer*, 253.
158 **"And then I suddenly"**: BL1, 554.
158 **"enfant terrible"**: Alvarez, *Incurable Physician*, 110.
158 **"I paid a price"**: Caroline Drewes, "Dr. Alvarez Still Talks Tough at 90," *San Francisco Examiner*, March 2, 1975.
159 **"In his own way"**: JRO, 932.
159 **"singularly unimpressed"**: BL2, 624.
160 **"I had been welcome"**: BL2, 692.
160 **"Ed McMillan and I were quite"**: HC, 14.
160 **"Some people"**: HC, 13.
160 **"He got to the point"**: HC, 11.
161 **"a few microseconds"**: Levy, "Alvarez the Nobelist," *California Monthly*, January/February 1969.
161 **"swindletron"**: Peter H. Rose, "The Tandem Accelerator," *DA*, 100.
161 **"the crackpot session"**: Glaser in Hargittai and Hargittai, *Candid Science VI*, 537.
162 **"a ball of junk"**: Galison, *Image and Logic*, 344.
163 **"I was trapped"**: Galison, 337.
164 **"but since the accelerators"**: "The Bubble Chamber Program at UCRL," April 18, 1955, B22, 1.
164 **"entire weapons system"**: Peter Galison, "Laboratory Life with Bubble Chambers," B35, 37.
165 **"One of the unknown"**: SAC (Los Angeles), April 16, 1951, LWA FBI file.

166 **"foo fighters"**: F. C. Durant, "Report of Meetings of Scientific Advisory Panel on Unidentified Flying Objects," January 14–18, 1953, 8.

166 **"not beyond"**: Durant, "Report of Meetings," 9.

166 **"gizmos"**: Wesley Price, "The Sky Is Haunted," *Saturday Evening Post*, March 6, 1948.

166 **"And I don't understand"**: Price, "Sky Is Haunted."

166 **"an outstanding scientist"**: Memo from Assistant Director, Scientific Intelligence, to Director of Central Intelligence, "Consultants for Advisory Panel on Unidentified Flying Objects," January 9, 1953.

166 **"our job"**: Page to James L. Klotz, October 3, 1992, Computer UFO Network, https://www.cufon.org/cufon/tp_corres.htm.

166 **"case histories"**: Durant, "Report of Meetings," 3.

167 **"interplanetary spacecraft"**: Ruppelt, *Report on Unidentified Flying Objects*, 210.

167 **"birds, balloons"**: Ruppelt, 220.

167 **"[He] thought"**: Ruppelt, 222.

167 **"peculiar radar echoes"**: Durant, "Report of Meetings," 18.

167 **"instrumental effects"**: Durant, 19.

168 **"with a little common"**: Clarke, *Astounding Days*, 88.

168 **"exciting days"**: LWA to Page, February 27, 1983, B47.

168 **"unexplained but not dangerous"**: Durant, "Report of Meetings," 9.

168 **"a direct physical threat"**: "Report of the Scientific Panel on Unidentified Flying Objects," January 17, 1953, 1.

168 **"a morbid national"**: "Report of the Scientific Panel," 2.

168 **"the aura"**: "Report of the Scientific Panel," 2.

168 **"I had the distinct"**: Vallee, *Forbidden Science*, 303.

169 **"scientists in gray"**: Hiltzik, *Big Science*, 411.

169 **"I sometimes wonder"**: Bird and Sherwin, *American Prometheus*, 559.

169 **"They like to be called"**: Bird and Sherwin, 557.

169 **"I knew that someone"**: BL2, 731.

170 **"People were anxious"**: LWA, "Excerpts from a Russian Diary," *Physics Today*, May 1957, 28.

170 **"He concluded"**: Segrè, *Mind Always in Motion*, 237.

170 **"Pontecorvo was very"**: LWA, "Further Excerpts from a Russian Diary," *Physics Today*, June 1957, 24.

170 **"I wouldn't be surprised"**: LWA, "Excerpts from a Russian Diary," 30.

171 **"Geraldine began spending"**: BL2, 677.

171 **"This wasn't a sudden"**: LWA to Gaither, March 28, 1957, Personal 1952–1964, C1.

172 **"I saw Jan every day"**: BL1, 14.

172 **"Have you seen"**: BL1, 14.

172 **"A divorce"**: BL2, 735.

173 **"extreme cruelty"**: "Dr. Alvarez Sued by Wife," *San Francisco Examiner*, November 26, 1957.

173 **"argued continuously"**: "Wife of Dr. Luis W. Alvarez Is Granted Divorce Here," *Oakland Tribune*, December 20, 1957.

173 **"It often occurs"**: Alvarez, *Incurable Physician*, 187–88.
173 **"It was his fault"**: Richard A. Muller interview, April 23, 2024.
174 **"Many people get married"**: Richard A. Muller interview, April 26, 2023.
174 **"I amused myself"**: BL2, 742.
175 **"electron beam weapon"**: Teller, *Memoirs*, 431.
175 **"one of the most fantastic"**: LWA to Richard Garwin, August 19, 1980, B27.
175 **"I would like to say"**: LWA to Wiesner, February 16, 1961, B9.
176 **"favorably impressed"**: LWA to William Shockley, November 22, 1961, B5.
177 **"interesting events"**: LWA, "Recent Developments in Particle Physics," *DA*, 134.
178 **"It would have been"**: Anne Roe interview, December 1962, AR, 8.
178 **"We were off"**: LWA, "Recent Developments in Particle Physics," *DA*, 137.
179 **"You are looking"**: Gerald M. Swatez, "The Social Organisation of a University Laboratory," *Minerva*, January 1970, 44.
179 **"[Lawrence] was gray"**: Hiltzik, *Big Science*, 390.
179 **"Luis was telling"**: Wallace Reynolds, quoted in Herken, *Brotherhood of the Bomb*, 412.
179 **"Angered when"**: Herken, *Brotherhood of the Bomb*, 321.
179 **"slow and painful death"**: LWA to Bogdan Maglich, March 1, 1976, B26.
179 **"Alvarez is brilliant"**: Herken, *Brotherhood of the Bomb*, chap. 19, note 73, accessed June 2024, https://brotherhoodofthebomb.com/bhbsource/endnotes.html.
180 **"Well, that leaves"**: BL1, "Hardball Physics," 27.
180 **"like a tyrant"**: BL2, 760.
181 **"His behavior"**: Galison, *Image and Logic*, 420.
181 **"bump hunting"**: LWA, "Recent Developments in Particle Physics," *DA*, 150.
182 **"the ideal way"**: Wojcicki, "My First Days in the Alvarez Group," *DA*, 170.
182 **"she was doing"**: "Alvarez Symposium—Lina Galtieri," YouTube video posted by Philip Dauber December 27, 2011, https://www.youtube.com/watch?v=wln14znX7rM.
182 **"in residence"**: "Report on Sabbatical Leave in Residence," Personal 1952–1964, C1.
182 **"tended to treat"**: Davis, *Lawrence and Oppenheimer*, 253.
182 **"The project business manager"**: Davis, 253.
182 **"like looking down"**: Paul Hernandez, quoted in Galison, *Image and Logic*, 315.
183 **"the person who talks"**: "Alvarez Symposium—Stan Wojcicki," YouTube video posted by Philip Dauber January 4, 2012, https://www.youtube.com/watch?v=wNrGSLqz8L4.
184 **"We'd have the answer"**: LWA, "Recent Developments in Particle Physics," *DA*, 131.
184 **"If you want to have"**: Swatez, "Social Organisation of a University Laboratory," 51.
184 **"prima donnas"**: Anne Roe interview, December 1962, AR, 25.
184 **"a rather major"**: LWA to Anne Roe, June 2, 1970, B10.
185 **"so foolhardy"**: BL1, 534.
185 **"a sensational picture"**: BL2, 744.

185 **"Many physicists felt"**: Kistiakowsky, *Scientist at the White House*, 143.
185 **"skimmed the cream"**: Swatez, "Social Organisation of a University Laboratory," 45.
185 **"a bad guy"**: Swatez, 56.
186 **"Most everybody"**: LWA to A. Abashian, November 13, 1980, B27.
186 **"I'll leave it"**: BL1, "Hardball Physics," 35.
186 **"I felt that I"**: LWA to Wiesner, February 16, 1961, B9.
186 **"the particle zoo"**: "100 Incredible Years of Physics—Particle Physics," Institute of Physics, accessed June 2024.
187 **"the unfortunate"**: BL2, 753.
187 **"there was no indication"**: BL2, 798.
187 **"Now things"**: Bird and Sherwin, *American Prometheus*, 576.

PART III

189 **"Richard Feynman"**: Rota, *Indiscrete Thoughts*, 202.

CHAPTER 7

191 **"I sensed a strong"**: Wojcicki, "My First Days in the Alvarez Group," *DA*, 167.
191 **"There is the famous"**: AIP2.
191 **"just a little dull"**: AIP2.
191 **"A physicist is no longer"**: LWA, "Recent Developments in Particle Physics," *DA*, 132.
191 **"beam shack"**: Swatez, "Social Organisation of a University Laboratory," *Minerva*, January 1970, 47.
191 **"They had merely"**: BL2, 21.
192 **"a real educational experience"**: BL2, 22.
192 **"[Students] ask fewer"**: Swatez, "Social Organisation of a University Laboratory," 47.
192 **"to become programmers"**: Swatez, 46.
192 **"If I were a graduate"**: AIP2.
192 **"I don't want"**: Klaw, *New Brahmins*, 147.
192 **"If a person"**: LWA to Anne Roe, June 2, 1970, B10.
192 **"unusually long"**: BL2, 799.
193 **"If the current"**: BL2, 895.
193 **"was about as long"**: LWA to Hornig, January 21, 1965, B9.
193 **"poet of the machine shop"**: Trower, *Luis W. Alvarez*, 12.
193 **"that my arms"**: "Remarks at the Convocation Ceremony—Pennsylvania College of Optometry," May 15, 1982, B47.
194 **"If it were a problem"**: Anne Roe interview, December 1962, AR, 18.
194 **"I knew there *was* a solution"**: LWA, "Development of Variable-Focus Lenses and a New Refractor," *DA*, 235.

194 **"phantom lenses":** LWA, "Development of Variable-Focus Lenses," 237.

194 **"to make a few million dollars":** Muller, *Nemesis*, 49.

195 **"on the dark side":** "Alvarez Symposium—Jack Lloyd 1," YouTube video posted by Philip Dauber December 27, 2011, https://www.youtube.com/watch?v=BvTZuZ5HEec.

198 **"They seemed to me":** LWA, "Physicist Examines the Kennedy Assassination Film," 215.

199 **"To me it means":** *A CBS News Inquiry: The Warren Report 1*, YouTube video posted by Conspiratard June 1, 2014, https://www.youtube.com/watch?v=XWtb2JwzkM8.

199 **"When we hear":** LWA, "Physicist Examines the Kennedy Assassination Film," 222.

199 **"That is why":** LWA, 222.

199 **"I don't feel":** LWA, 216.

199 **"the recollections":** LWA, quoted in Thompson, *Last Second in Dallas*, 120.

199 **"I observed hair":** Thompson, 402.

200 **"All witnesses":** LWA to Norman Ramsey, November 16, 1981, B49.

200 **"What that guy":** Zapruder, *Twenty-Six Seconds*, 202.

202 **"Why would Chephren":** LWA, "A Proposal to 'X-Ray' the Egyptian Pyramids to Search for Presently Unknown Chambers," March 1, 1965, Berkeley Physics Memo 544, 2.

202 **"just as a rifle":** LWA, "Proposal to 'X-Ray' the Egyptian Pyramids," 8.

203 **"Such a wrong":** Bird and Sherwin, *American Prometheus*, 4.

205 **"a dirty spying":** BL2, 816.

205 **"I think you":** Yazolino to LWA, September 21, 1968, B30.

205 **"There is no doubt":** Telegram from Yazolino to LWA, September 22, 1968, B30.

206 **"In addition":** LWA, "Using Cosmic Rays in the Search for Hidden Chambers in the Pyramids," *DA*, 183.

207 **"I agreed":** BL2, 757.

207 **"I assumed":** BL2, 758.

207 **"Every high-energy":** BL2, 760.

208 **"equally weak":** BL2, 723.

208 **"I had a visit":** BL2, 758.

208 **"the implacable":** Segrè, *Mind Always in Motion*, 267.

208 **"total flops":** BL2, 762.

208 **"I do feel":** LWA to McMillan, July 17, 1963, B32.

208 **"so that Ed":** BL2, 762.

208 **"passed its prime":** LWA to McMillan, March 18, 1968, B9.

210 **"You're kidding":** Kay Wahl, "Nobel Prize to U.C.'s Alvarez," *Oakland Tribune*, October 30, 1968.

210 **"Have you seen":** Rosenfeld in "Alvarez Symposium—After Dinner Part 2," YouTube video posted by Philip Dauber January 6, 2012, https://www.youtube.com/watch?v=QqECBByYHW0.

210 **"for your decisive":** *DA*, photo section following 202.

210 **"I don't attribute"**: BL2, 767.
211 **"unlike many other"**: Erik Rydberg, quoted in Wahl, "Nobel Prize to U.C.'s Alvarez."
211 **"Ernest had told"**: BL2, 772.
211 **"to see no men"**: BL2, 780.
212 **"If we spend"**: LWA, "Recent Developments in Particle Physics," *DA*, 110.
212 **"blatant exercise"**: BL2, 761.
212 **"I purposely refrained"**: BL1, "Hardball Physics," 24.
212 **"Students and colleagues"**: Associated Press, "The Nobel—How It Can Affect Lives," *Oakland Tribune*, February 2, 1977.
212 **"swift disengagement"**: "43 Nobel Winners Ask End to War," *Boston Globe*, June 4, 1970.
212 **"very exciting signals"**: LWA et al., "Search for Hidden Chambers in the Pyramids," *DA*, 192.
212 **"If it refuses"**: LWA to Seaborg, March 22, 1969, B10.
213 **"failure"**: "Tomb Hunt in Pyramid a Failure," *Oakland Tribune*, April 30, 1969.
213 **"I have never"**: National Geographic Society, "Pharaoh Keeps His Secrets," *Alabama Journal* (Montgomery, AL), May 9, 1969.
213 **"If we pull out"**: United Press International, "The Advance of Science," *York (PA) Dispatch*, May 29, 1969.
213 **"pyramidiots"**: LWA to G. Legman, January 28, 1971, B32.
213 **"There is some force"**: Amr Goneid, quoted by Tompkins, *Secrets of the Great Pyramid*, 275 (originally in John Tunstall, "Pyramid versus the Space Age," *London Times*, July 20, 1969).
213 **"nut file"**: Rubén Martínez, "Adventures of Luis Alvarez," 170–234.
214 **"interesting material"**: LWA to Shockley, July 9, 1969, B10.
214 **"comes on pretty strong"**: LWA to Clarence Mayhew, February 11, 1970, Bohemian Club 1951–1985, C1.
214 **"You have to get"**: Carl Irving, "Physicists Clash on Atom Fund," *Oakland Tribune*, June 25, 1963.
214 **"one of the great"**: Galison, *Image and Logic*, 422.
215 **"We shall therefore"**: LWA, "Study of High-Energy Interactions Using a 'Beam' of Primary Cosmic Ray Protons," May 4, 1964, Berkeley Physics Memo 503, 4.
215 **"missing element"**: LWA, quoted in Levy, "Alvarez the Nobelist," *California Monthly*, January/February 1969.
215 **"Young graduate students"**: AIP2.
216 **"We never found"**: Richard A. Muller, "Did My 1967 Experiment Inspire China's Spy Balloon?," *Wall Street Journal*, February 17, 2023.
216 **"fiascoes"**: LWA to Walter Orr Roberts, July 21, 1970, B10.
217 **"It also, in the end"**: Smoot and Davidson, *Wrinkles in Time*, 88.
217 **"enriched sample"**: A. Buffington et al., "Search for Antimatter in Primary Cosmic Rays," *DA*, 201.
218 **"Let's be sure"**: Andrew Buffington, "Looking for Antimatter in the Cosmic Rays," *DA*, 196.

CHAPTER 8

219 **"With the galactic"**: LWA, "A Possible Explanation of High Energy Cosmic Ray Phenomena in Terms of Dirac Magnetic Monopoles," November 7, 1969, Berkeley Physics Memo 479, 5.

220 **"One would be surprised"**: Dirac, "Quantised Singularities in the Electromagnetic Field," *Proceedings of the Royal Society* Series A, 133 (September 1931): 71.

221 **"All were smiling"**: Lincoln Kilian, "In Search of the Elusive Monopole," *San Francisco Sunday Examiner & Chronicle*, December 5, 1982.

222 **"after voting"**: BL2, 693.

222 **"shabby, disgusting"**: "Weighing In on the Watergate Tapes," *The New Yorker*, May 12, 1974.

222 **"One can't be"**: LWA China diary, January 12, 1973, B71.

222 **"in residence"**: LWA to Eugene Commins, March 12, 1974, B71.

223 **"I have an intense"**: "The Early Days of Accelerator Mass Spectrometry," B33, 2.

223 **"We remember"**: Robert D. Watt et al., "Luis W. Alvarez," University of California: In Memoriam, 1988, accessed June 2024, http://texts.cdlib.org/view?docId=hb967nb5k3.

224 **"Their goals"**: Trower, *Luis W. Alvarez*, 13.

224 **"a hamburger"**: "Alvarez Symposium—Jack Lloyd 1," YouTube video posted by Philip Dauber December 27, 2011, https://www.youtube.com/watch?v=BvTZuZ5HEec.

224 **"I used to read"**: Anne Roe interview (ca. 1952), AR, 4.

224 **"If you can't think up"**: Roe interview, 7.

224 **"I had long wondered"**: BL2, 899.

225 **"I considered"**: LWA to Mrs. Lewis Strauss (Alice), January 24, 1974, B20.

225 **"one of the major"**: "Bring It Back Alive," *Time*, August 25, 1975.

226 **"going into battle"**: Don Alvarez interview, April 24, 2023.

226 **"completely dissatisfied"**: "Memorandum of a Discussion Concerning the Monopole Paper by Price and Shirk," August 22, 1975, B55.

227 **"as vigorously"**: LWA, "Analysis of a Reported Magnetic Monopole," August 27, 1975, LBL-4260, 4.

227 **"What otherwise"**: LWA, "Analysis of a Reported Magnetic Monopole," 17.

227 **"strictly not cricket"**: Madlyn Resener, "Professor Bounces Back after Monopole Miss," *North East Bay Independent and Gazette* (Berkeley, CA), November 27, 1978.

227 **"a bit strained"**: LWA to James Barger, July 8, 1981, B33.

227 **"Let me warn"**: Dewey McLean, "The Rancorous Asteroid Impact vs. Deccan Traps Volcanism Dinosaur Extinction Debate," The Deccan Traps Volcanism-Greenhouse Dinosaur Extinction Theory, accessed June 2024, https://web.archive.org/web/20140402205237/http://filebox.vt.edu/artsci/geology/mclean/Dinosaur_Volcano_Extinction/pages/scienpol.html.

227 **"young graduate students"**: AIP2.

227 **"When can you start?"**: Muller, *Nemesis*, 21.

228 **"Just go over there"**: Muller, 23.

228 **"Now I know"**: Muller, 24.

228 **"certainly the best"**: LWA introduction to Muller, *Nemesis*, xi.
228 **"[Alvarez] had learned"**: Muller, *Nemesis*, 27.
228 **"he referred"**: Richard A. Muller interview, April 26, 2023.
228 **"to think crazy"**: Muller, *Now*, 150.
228 **"Only one out of ten"**: "Alvarez Symposium—Rich Muller," YouTube video posted by Philip Dauber December 28, 2011, https://www.youtube.com/watch?v=_5dDUjoZMe4.
229 **"Have you solved it?"**: Richard A. Muller interview, April 26, 2023.
229 **"nothing looks"**: LWA, textbook draft, C3, II-17.
229 **"Before you rush"**: Smoot and Davidson, *Wrinkles in Time*, 113.
230 **"Just as more water"**: Muller, *Now*, 143.
230 **"a waste of time"**: "Alvarez Symposium—Rich Muller."
230 **"Wow, that's really"**: "Alvarez Symposium—Rich Muller."
231 **"Luie was considered"**: Mather and Boslough, *Very First Light*, 125.
232 **"Luie thought"**: Quotes on the integral quark search are from Richard A. Muller, "Submarines, Quarks, and Radioisotope Dating," *DA*, 226–28.
234 **"If you ever knew"**: "Alvarez Symposium—Carl Pennypacker," YouTube video posted by Philip Dauber December 28, 2011, https://www.youtube.com/watch?v=A57CVoP59Fw.
234 **"If your junk pile"**: "Alvarez Symposium—Carl Pennypacker."
234 **"to escape from"**: LWA to William Nierenberg, December 2, 1985, B47.
234 **"There was indication"**: Carter, *White House Diary*, 357.
235 **"statistically insignificant"**: BL1, "Scientific Detective Work," 34.
236 **"reasonable"**: "Ad Hoc Panel on the September 22 Event," May 23, 1980, 18.
236 **"was probably not"**: "Ad Hoc Panel on the September 22 Event," 19.
236 **"very unusual"**: George N. Oetzel and Steven C. Johnson, *Vela Meteoroid Evaluation*, SRI International, January 29, 1980, 44.
236 **"phase anomaly"**: Leonard Weiss, "Flash from the Past," *Bulletin of the Atomic Scientists*, September 8, 2015.
236 **"had been seen"**: Weiss, "Flash from the Past."
237 **"fuck-up"**: Hersh, *Samson Option*, 271–72.
237 **"My job"**: Avner Cohen and William Burr, "Revisiting the 1979 VELA Mystery," *Sources and Methods* (blog), History and Public Policy Program, Wilson Center, August 31, 2020.
237 **"uninstructable"**: Panofsky, *Physics, Politics, and Peace*, 160.
237 **"People who know"**: Cohen and Burr, "Revisiting the 1979 VELA Mystery."
237 **"overcame his usual"**: Rhodes, *Twilight of the Bombs*, 167.
237 **"I have no further"**: LWA to Raymond Marcus, September 12, 1967, B51.
238 **"to persuade"**: Thompson, *Last Second in Dallas*, 119.
238 **"I once leafed"**: LWA to Walter Menaker, August 15, 1968, B51.
239 **"when I couldn't"**: LWA handwritten notes on jet effect, provided to the author by Paul Hoch. See also Thompson, *Six Seconds in Dallas*, 91.
239 **"Whether a melon"**: Thompson, *Last Second in Dallas*, 125.
239 **"was apparently important"**: LWA, "Physicist Examines the Kennedy Assassination Film," 224.

239 **"would have struck"**: Thompson, *Last Second in Dallas*, 126.
240 **"a documented counterexample"**: LWA, "Physicist Examines the Kennedy Assassination Film," 219.
240 **"The deposit"**: "Investigation of the Assassination of President John F. Kennedy," Hearings Before the Select Committee on Assassinations, September 6–9, 1978, vol. 1, 414.
241 **"impulse sounds"**: Barger in "Investigation of the Assassination," 18.
241 **"about even"**: Barger in "Investigation of the Assassination," 94.
241 **"very similar"**: Bugliosi, *Reclaiming History*, endnotes, 167.
242 **"scientific acoustical evidence"**: "JFK Assassination Records—Findings," National Archives, accessed June 2024, https://www.archives.gov/research/jfk/select-committee-report/part-1b.html.
242 **"probably assassinated"**: "JFK Assassination Records—Findings," https://www.archives.gov/research/jfk/select-committee-report/part-1c.html.
242 **"particularly shoddy"**: LWA to Norman Ramsey, October 20, 1980, B49.
242 **"damning it"**: LWA to Ramsey, October 20, 1980.
242 **"evidence that destroyed"**: Thompson, *Last Second in Dallas*, 294.
243 **"Hold everything secure"**: Thompson, 290.
243 **"much time"**: Barber to "Sirs," October 6, 1980, B49.
243 **"I think [Barber]"**: C. K. Reed to LWA, November 17, 1980, B49.
243 **"Chaney"**: Ramsey to James Barger, January 26, 1981, B49.
244 **"He didn't care"**: Thompson, *Last Second in Dallas*, 433.
244 **"I'll check it"**: Norman Ramsey to Jeff Fogel, March 3, 1981, B49.
244 **"I'm sorry"**: Bowles to Norman Ramsey, November 24, 1981, B49.
244 **"discounted as being"**: Bowles to Norman Ramsey, December 30, 1981, B49.
245 **"discounted by sound spectrograms"**: "Report of the Committee on Ballistic Acoustics," 1982, 71.
245 **"grossly misused"**: Robert Phinney to Norman Ramsey, April 16, 1981, B50.
245 **"all three networks"**: LWA to Ramsey Panel members, January 13, 1982, B50.
245 **"reliable acoustic data"**: "Report of the Committee on Ballistic Acoustics," 34.

CHAPTER 9

247 **"Dad did not"**: Alvarez, *T. rex and the Crater of Doom*, 60.
248 **"Luis Alvarez freely"**: Muller, *Nemesis*, 40.
249 **"about the rate"**: Alvarez, *T. rex and the Crater of Doom*, 56.
250 **"Choosing what problems"**: Alvarez, 42.
250 **"For some reason"**: Quotes on Muller and Walter Alvarez's attempted collaboration are from Muller, *Nemesis*, 33–38.
252 **"There is every reason"**: LWA to Walter Alvarez, May 6, 1977, B71.
253 **"And I think"**: Muller, *Nemesis*, 40.
253 **"Maybe geology"**: Richard Rhodes interview, February 22, 2023.
253 **"He had always"**: Muller, *Nemesis*, 40.
253 **"have the fun"**: LWA to Walter Alvarez, May 6, 1977, B71.

253 **"He had made so many"**: Muller, *Nemesis*, 42.
254 **"the intellectual heir"**: Alvarez, *T. rex and the Crater of Doom*, 67.
254 **"seriously wrong"**: "Alvarez Symposium—Walter Alvarez," YouTube video posted by Philip Dauber December 26, 2011, https://www.youtube.com/watch?v=YR31ctMJM3c.
254 **"We were happy"**: Asaro, "The Cretaceous–Tertiary Iridium Anomaly and the Asteroid Impact Theory," *DA*, 241.
254 **"bump hunter"**: Wilford, *Riddle of the Dinosaur*, 223.
254 **"It's the most exciting"**: "NOVA—The Asteroid and the Dinosaur (1981)," YouTube video posted by Dinosaur Documentaries October 30, 2022, https://www.youtube.com/watch?v=RBnW1u20jWQ.
254 **"like a shark"**: Muller, *Nemesis*, 51.
255 **"Meet me"**: Muller, 56.
256 **"It was very rare"**: Muller, 56.
256 **"When you have a theory"**: Richard A. Muller interview, April 26, 2023.
256 **"A solar flare"**: Marcy Kates, "Why the Dinosaurs Died—A New Theory," *North East Bay Independent and Gazette* (Berkeley, CA), May 28, 1979.
256 **"Repeat every single"**: Alvarez, *T. rex and the Crater of Doom*, 74.
256 **"There are several"**: Muller, *Nemesis*, 59.
257 **"He told everyone"**: Muller, 61.
258 **"twilight glows"**: Archibald in Symons, *Eruption of Krakatoa*, 426.
258 **"into the loftier"**: Symons, 452.
258 **"After two weeks"**: Muller, *Nemesis*, 66.
258 **"would be delighted"**: Alvarez, *T. rex and the Crater of Doom*, 78.
259 **"until the son arrived"**: Richard A. Muller interview, April 26, 2023.
259 **"Now it was too late"**: Muller, *Nemesis*, 67.
259 **"You are right"**: Muller, 68.
260 **"And at least N-1"**: Walter Alvarez email, June 2, 2024.
260 **"maverick ideas"**: Raup, *Nemesis Affair*, 42.
260 **"The resulting global"**: Mike Toner, "Dinosaurs May Have Gone Out with a Bang," *Miami Herald*, January 6, 1980.
260 **"on nuts"**: George Alexander, "Asteroid Did In Dinosaurs, UC Scientists Theorize," *Los Angeles Times*, January 5, 1980.
260 **"a humming telephone"**: LWA to Ken Hsü, February 6, 1980, B35.
260 **"too quick to call"**: Raup, *Nemesis Affair*, 71.
261 **"And so was born"**: Wilford, *Riddle of the Dinosaur*, 227.
261 **"we would be wasting"**: LWA to David Raup, December 14, 1981, B27.
262 **"just three meters"**: LWA to Raup.
262 **"the evolutionary history"**: Clemens in Glen, *Mass-Extinction Debates*, 239.
262 **"Different groups"**: Raup, *Nemesis Affair*, 72.
262 **"the biosphere"**: LWA to Stephen Jay Gould, February 10, 1983, B47.
263 **"Scientists have known"**: Margaret Munro, "Scientist Predicts Earth Calamity if Man Doesn't Track Asteroids," *Ottawa Citizen*, May 22, 1981.
263 **"complicated"**: Archibald, *Dinosaur Extinction and the End of an Era*, 46.
264 **"dragged, kicking"**: LWA to Dale Russell, April 7, 1981, B27.

264 **"crazies"**: LWA to Thomas Budinger, August 28, 1982, B47.
264 **"Many scientists"**: LWA to Daniel Koshland, March 1, 1985, B47.
264 **"had all the characteristics"**: Muller, *Nemesis*, 70.
264 **"And yet in the fifteen"**: Muller, 73.
264 **"bad genes"**: Raup, *Nemesis Affair*, 73.
264 **"idiosyncratic preference"**: Gould, *Dinosaur in a Haystack*, 152.
265 **"Always look"**: Gould, *Structure of Evolutionary Theory*, 1308.
265 **"codswallop"**: Muller, *Nemesis*, 72.
265 **"It is rare"**: Muller, 80.
265 **"glared red-faced"**: Dewey McLean, "The Rancorous Asteroid Impact vs. Deccan Traps Volcanism Dinosaur Extinction Debate," The Deccan Traps Volcanism-Greenhouse Dinosaur Extinction Theory, accessed June 2024, https://web.archive.org/web/20140402205237/http://filebox.vt.edu/artsci/geology/mclean/Dinosaur_Volcano_Extinction/pages/scienpol.html.
265 **"Do you plan"**: McLean.
266 **"heated exchanges"**: Alvarez, *T-rex and the Crater of Doom*, 99.
266 **"a widespread virtual vendetta"**: McLean, "Rancorous Asteroid Impact."
266 **"As a horrible"**: LWA to Jastrow, January 4, 1984, B47.
266 **"All I can say"**: Munro, "Scientist Predicts Earth Calamity."
267 **"death from the heavens"**: McLean, "Rancorous Asteroid Impact."
267 **"The discovery"**: R. P. Turco et al., "Nuclear Winter: Global Consequences of Multiple Nuclear Explosions," *Science*, December 1983, 1283.
268 **"There seems to be a real"**: Carl Sagan, "Nuclear Winter," The School of Cooperative Individualism, accessed June 2024, https://www.cooperative-individualism.org/sagan-carl_nuclear-winter-1983.htm.
268 **"*almost* the last"**: LWA to David Inouye, November 6, 1980, B27.
268 **"That the asteroid"**: LWA, "Experimental Evidence That an Asteroid Impact Led to the Extinction of Many Species 65 Million Years Ago," in Metropolis, *New Directions in Physics*, 37.
268 **"instantaneously"**: Metropolis, 43.
269 **"arrogance"**: Malcolm W. Browne, "Dinosaur Experts Resist Meteor Extinction Idea," *New York Times*, October 29, 1985.
269 **"to sit down"**: Bill Manson, "Beastly Debate," *Los Angeles Times*, March 24, 1988.
269 **"Terrestrial events"**: "Miscasting the Dinosaur's Horoscope," *New York Times*, April 2, 1985.
269 **"would editorialize"**: Gould, *Structure of Evolutionary Theory*, 1303.
269 **"an April Fool's joke"**: Richard A. Muller and Walter Alvarez, "Was It Nemesis That Killed the Dinosaur?," *New York Times*, April 14, 1985.
270 **"concern among"**: Muller, *Nemesis*, 84.
270 **"all but a few diehards"**: Natalie Angier, "Did Comets Kill the Dinosaurs?," *Time*, May 6, 1985.
270 **"probable enough"**: LWA to Frank Asaro, Helen Michel, Walter Alvarez, July 14, 1979, B26.
270 **"the same could be true"**: LWA, Walter Alvarez, Frank Asaro, and Helen V. Michel, "Extraterrestrial Cause for the Cretaceous–Tertiary Extinction," *DA*, 255.

271 **"cheating"**: Muller, *Nemesis*, 6.
271 **"Suppose someday"**: Muller, 7–9.
272 **"I really hope"**: LWA to Raup, October 11, 1983, B47.
272 **"A few months"**: Muller, *Nemesis*, 98.
272 **"To be a really good"**: Muller, 103.
272 **"Of course, the star"**: Muller, 104.
273 **"I had the feeling"**: Muller, 111.
273 **"Because that is where"**: Crater-discussion quotes are from Muller, *Nemesis*, 120–32.
274 **"Nemesis is right now"**: Muller, *Nemesis*, 143.
275 **"The real effect"**: Muller, 144.
275 **"To exaggerate"**: Raup, *Nemesis Affair*, 154.
275 **"Perhaps periodic"**: Muller, *Nemesis*, 154.
276 **"the most exciting"**: Muller, 158.
276 **"finding needles"**: Hargittai and Hargittai, *Candid Science V*, 217.
276 **"nitrogen camera"**: Trower, *Luis W. Alvarez*, 17.
276 **"unpublishable"**: Rubén Martínez, "Adventures of Luis Alvarez," 57.
276 **"I don't want"**: Richard Rhodes interview, February 22, 2023.
277 **"Don't try"**: Stoll, *Cuckoo's Egg*, 92–93.
277 **"From that grew"**: Stoll in "Alvarez Symposium—After Dinner Part 1," YouTube video posted by Philip Dauber January 6, 2012, https://www.youtube.com/watch?v=xnqcv_Jzhgo.
277 **"I don't like"**: Malcolm W. Browne, "The Debate over Dinosaur Extinctions Takes an Unusually Rancorous Turn," *New York Times*, January 19, 1988.
278 **"All science"**: Garson O'Toole, "All Science Is Either Physics or Stamp Collecting," Quote Investigator, May 8, 2015, https://quoteinvestigator.com/2015/05/08/stamp.
278 **"inept"**: Browne, "Debate over Dinosaur Extinctions."
278 **"psychotic"**: Richard Sandomir, "What's All This Fuss about Dinosaurs?," *Newsday*, May 1, 1988.
278 **"This is my last"**: Browne, "Debate over Dinosaur Extinctions."
278 **"So you might as well"**: LWA to "Friends," March 1988, B48.
278 **"I thanked them"**: LWA to "Friends."
279 **"I am weaker"**: LWA to "Friends."
279 **"The surgeries"**: Trower, *Luis W. Alvarez*, 18.
280 **"A few days later"**: Trower, 18.
280 **"a prince of science"**: Gould, *Wonderful Life*, 281.
280 **"My father"**: Powell, *Night Comes to the Cretaceous*, 165.

EPILOGUE

281 **"So complex"**: Jorge Luis Borges, "On William Beckford's *Vathek*," Eliot Weinberger, trans., in *Selected Non-Fictions*, ed. Eliot Weinberger (New York: Viking, 1999).

283 **"the late Cretaceous extinctions"**: Powell, *Night Comes to the Cretaceous*, 103.
284 **"the smoking gun"**: Alvarez, *T. rex and the Crater of Doom*, 123.
284 **"Those eleven years"**: Kolbert, *Sixth Extinction*, 82.
284 **"dinosaur in a haystack"**: Gould, *Dinosaur in a Haystack*, 147.
284 **"The Alvarez theory"**: Gould, 153.
285 **"If twenty-five years"**: Kolbert, *Sixth Extinction*, 104.
285 **"disappointed and disturbed"**: Glen, *Mass-Extinction Debates*, 267.
285 **"To learn that he"**: Gould, *Structure of Evolutionary Theory*, 1131.
286 **"the most important"**: Smoot and Davidson, *Wrinkles in Time*, 283.
286 **"I had learned"**: Muller, *Now*, 154.
286 **"One of the things"**: "Alvarez Symposium—Saul Perlmutter," YouTube video posted by Philip Dauber December 28, 2011, https://www.youtube.com/watch?v=XG7tIs5HYiE.
286 **"cowboy spirit"**: "Alvarez Symposium—Saul Perlmutter."
287 **"I wish we had worked"**: Amina Khan, "Using Particle Physics, Scientists Find Hidden Structure inside Egypt's Great Pyramid," *San Diego Union-Tribune*, November 2, 2017.
287 **"I didn't ask Luie"**: Rhodes, *Hole in the World*, 78.
288 **"So long as the Jap"**: "A Jap Burns," *Life*, August 13, 1945, 34.
288 **"the physicists have known sin"**: BL2, 582.
288 **"one of the most lifesaving"**: BL2, 587.
288 **"I can think of nothing"**: BL2, 606.
288 **"hydrogen bombs"**: BL2, 345.
288 **"I have frequently"**: BL2, 695.
288 **"as an acknowledged"**: BL2, 603.
289 **"When we physicists"**: LWA, "Adventures in Nuclear Physics" (Faculty Research Lecture), March 1962, 32–34.
289 **"We know the basic"**: Cornelius A. Tobias, "Heavy Ions," *DA*, 52.
289 **"Most people have no"**: BL2, 605.
290 **"severe climactic changes"**: LWA to Gordon Allen, September 24, 1979.
290 **"It becomes morally"**: Horst W. J. Rittel and Melvin M. Webber, "Dilemmas in a General Theory of Planning," *Policy Sciences* 4 (1973): 161.
291 **"The social professions"**: Rittel and Webber, "Dilemmas," 160.
291 **"a single experimenter"**: BL2, 152.
291 **"stimulated my imagination"**: BL2, 152.
292 **"He contrasted trying"**: Wojcicki, "My First Days in the Alvarez Group," *DA*, 165.
292 **"because of the interpersonal"**: BL2, 901.
292 **"I brought the computers in"**: AIP2.
292 **"[A student] ends up"**: LWA to H. H. Staub, February 8, 1974, B20.
293 **"crazy experiments"**: BL2, 142.
293 **"in more fields"**: BL2, 907.
294 **"tricks of the trade"**: Rubén Martínez, "Adventures of Luis Alvarez," 126.
294 **"only through informal"**: LWA, textbook draft, C3, III-1.
294 **"You don't want"**: AIP2.

295 **"Learning how to pick"**: Muller, *Nemesis*, 160–61.
295 **"The fox knows"**: Berlin, *Hedgehog and the Fox*, 1.
295 **"lead lives"**: Berlin.
295 **"Smoot is content"**: LWA to Gerald Freund, February 16, 1982, B47.
295 **"Creativity in science"**: LWA to Anne Roe, August 13, 1968, B10.

BIBLIOGRAPHY

Albright, Joseph, and Marcia Kunstel. *Bombshell: The Secret Story of America's Unknown Atomic Spy Conspiracy*. New York: Times Books, 1997.

Alvarez, Luis W. *Alfred Lee Loomis, 1887–1975*. Washington, DC: National Academy of Sciences, 1980.

——. *Alvarez: Adventures of a Physicist*. New York: Basic Books, 1987.

——. *Discovering Alvarez: Selected Works of Luis W. Alvarez, with Commentary by His Students and Colleagues*. Edited by W. Peter Trower. Chicago: University of Chicago Press, 1987.

——. *Ernest Orlando Lawrence, 1901–1958*. Washington, DC: National Academy of Sciences, 1970.

Alvarez, Walter. *T-rex and the Crater of Doom*. Princeton: Princeton University Press, 1997.

Alvarez, Walter C. *Incurable Physician: An Autobiography*. Englewood Cliffs, NJ: Prentice-Hall, 1963.

Archibald, J. David. *Dinosaur Extinction and the End of an Era*. New York: Columbia University Press, 1996.

Berlin, Isaiah. *The Hedgehog and the Fox: An Essay on Tolstoy's Philosophy of History*. New York: Simon & Schuster, 1953.

Bird, Kai, and Martin Sherwin. *American Prometheus: The Triumph and Tragedy of J. Robert Oppenheimer*. New York: Vintage Books, 2006.

Buderi, Robert. *The Invention That Changed the World: How a Small Group of Radar Pioneers Won the Second World War and Launched a Technological Revolution*. New York: Simon & Schuster, 1996.

Bugliosi, Vincent. *Reclaiming History: The Assassination of President John F. Kennedy*. New York: W. W. Norton, 2007.

Carter, Jimmy. *White House Diary*. New York: Farrar, Straus and Giroux, 2010.

Childs, Herbert. *An American Genius: The Life of Ernest Orlando Lawrence, Father of the Cyclotron*. New York: E. P. Dutton, 1968.

Clarke, Arthur C. *Astounding Days*. New York: Bantam, 1989.

——. *Glide Path*. New York: Dell, 1963.

Cole, K. C. *Something Incredibly Wonderful Happens: Frank Oppenheimer and His Astonishing Exploratorium*. Chicago: University of Chicago Press, 2009.

Conant, Jennet. *Tuxedo Park: A Wall Street Tycoon and the Secret Palace of Science That Changed the Course of World War II*. New York: Simon & Schuster, 2002.

Coster-Mullen, John. *Atom Bombs*. Self-published, 2020.

Davis, Nuel Pharr. *Lawrence and Oppenheimer*. New York: Simon & Schuster, 1968.

Feynman, Richard, as told to Ralph Leighton. *What Do You Care What Other People Think?* New York: W. W. Norton, 1988.

Five Years at the Radiation Laboratory. Cambridge, MA: Massachusetts Institute of Technology, 1946.

Galison, Peter. *Image and Logic: A Material Culture of Microphysics.* Chicago: University of Chicago Press, 1997.

Glen, William, ed. *The Mass-Extinction Debates.* Redwood City, CA: Stanford University Press, 1994.

Gould, Stephen Jay. *Dinosaur in a Haystack.* New York: Harmony, 1995.

———. *The Structure of Evolutionary Theory.* Cambridge, MA: The Belknap Press of Harvard University, 2002.

———. *Wonderful Life.* New York: W. W. Norton, 1989.

Greenberg, Daniel S. *The Politics of Pure Science.* Chicago: University of Chicago Press, 1999.

Greenstein, George. *Portraits of Discovery.* New York: John Wiley & Sons, 1998.

Hargittai, Balazs, and István Hargittai. *Candid Science V: Conversations with Famous Scientists.* London: Imperial College Press, 2005.

———. *Candid Science VI.* London: Imperial College Press, 2006.

Heilbron, J. L., and Robert W. Seidel. *Lawrence and His Laboratory.* Berkeley: University of California Press, 1989.

Herken, Gregg. *Brotherhood of the Bomb: The Tangled Lives and Loyalties of Robert Oppenheimer, Ernest Lawrence, and Edward Teller.* New York: Henry Holt, 2002.

Hersh, Seymour. *The Samson Option: Israel's Nuclear Arsenal and American Foreign Policy.* New York: Random House, 1991.

Hiltzik, Michael. *Big Science: Ernest Lawrence and the Invention That Launched the Military-Industrial Complex.* New York: Simon & Schuster, 2015.

Hoddeson, Lillian, Paul W. Henriksen, Roger A. Meade, and Catherine Westfall. *Critical Assembly: A Technical History of Los Alamos during the Oppenheimer Years.* Cambridge: Cambridge University Press, 1993.

Hope, Bob. *I Never Left Home.* New York: Simon & Schuster, 1944.

Jungk, Robert. *Brighter Than a Thousand Suns: A Personal History of the Atomic Scientists.* Translated by James Cleugh. New York: Harcourt, Brace, 1958.

Kistiakowsky, George B. *A Scientist at the White House.* Cambridge, MA: Harvard University Press, 1976.

Klaw, Spencer. *The New Brahmins: Scientific Life in America.* New York: William Morrow, 1968.

Knebel, Fletcher, and Charles W. Bailey II. *No High Ground.* Westport, CT: Greenwood, 1960.

Kolbert, Elizabeth. *The Sixth Extinction.* New York: Henry Holt, 2014.

Libby, Leona Marshall. *The Uranium People.* New York: Crane Russak, 1979.

Lilienthal, David E. *The Journals of David E. Lilienthal.* Vol. 2, *The Atomic Energy Years, 1945–1950.* New York: Harper & Row, 1964.

Mather, John C., and John Boslough. *The Very First Light: The True Inside Story of the Scientific Journey Back to the Dawn of the Universe.* New York: Basic Books, 2008.

McMillan, Priscilla J. *The Ruin of J. Robert Oppenheimer and the Birth of the Modern Arms Race*. New York: Viking, 2005.

Metropolis, Nicholas, ed. *New Directions in Physics*. Boston: Academic Press, 1987.

Muller, Richard A. *Nemesis: The Death Star*. New York: Weidenfeld & Nicholson, 1988.

——. *Now: The Physics of Time*. New York: W. W. Norton, 2016.

Panofsky, Wolfgang K. H. *Panofsky on Physics, Politics, and Peace*. New York: Springer, 2007.

Powell, James Lawrence. *Night Comes to the Cretaceous*. New York: W. H. Freeman, 1998.

Raup, David M. *The Nemesis Affair: A Story of the Death of the Dinosaurs and the Ways of Science*. New York: W. W. Norton, 1999.

Rhodes, Richard. *Dark Sun: The Making of the Hydrogen Bomb*. New York: Simon & Schuster, 1995.

——. *A Hole in the World*. Lawrence: University Press of Kansas, 2000.

——. *The Making of the Atomic Bomb*. New York: Simon & Schuster, 1986.

——. *The Twilight of the Bombs*. New York: Alfred A. Knopf, 2010.

Richelson, Jeffrey T. *Spying on the Bomb*. New York: W. W. Norton, 2006.

Roberts, Sam. *The Brother: The Untold Story of the Rosenberg Case*. New York: Random House, 2001.

Roe, Anne. *The Making of a Scientist*. New York: Dodd, Mead, 1953.

Rota, Gian-Carlo. *Indiscrete Thoughts*. Boston: Birkhäuser, 1997.

Rubén Martínez, Jesús. "The Adventures of Luis Alvarez." PhD diss., University of Texas at Austin, 2011.

Ruppelt, Edward J. *The Report on Unidentified Flying Objects*. Garden City, NY: Doubleday, 1956.

Schwartz, David N. *The Last Man Who Knew Everything: The Life and Times of Enrico Fermi, Father of the Nuclear Age*. New York: Basic Books, 2017.

Seaborg, Glenn T., with Eric Seaborg. *Adventures in the Atomic Age*. New York: Farrar, Strauss and Giroux, 2001.

Segrè, Emilio. *A Mind Always in Motion*. Berkeley: University of California Press, 1993.

Serber, Robert, with Robert P. Crease. *Peace and War*. New York: Columbia University Press, 1998.

Slosson, Edwin E. *Creative Chemistry*. London: University of London Press, 1921.

Smoot, George, and Keay Davidson. *Wrinkles in Time*. New York: William Morrow, 1993.

Solberg, Carl. *Conquest of the Skies*. Boston: Little, Brown, 1979.

Stern, Philip M., with Harold P. Green. *The Oppenheimer Case*. New York: Harper & Row, 1969.

Stoll, Cliff. *The Cuckoo's Egg: Tracking a Spy through the Maze of Computer Espionage*. New York: Pocket Books, 1990.

Symons, G. J., ed. *The Eruption of Krakatoa, and Subsequent Phenomena*. London: Harrison and Sons, 1888.

Teller, Edward. *Memoirs*. Cambridge, MA: Perseus, 2001.

Thomas, Gordon, and Max Morgan Witts. *Enola Gay*. New York: Stein and Day, 1977.

Thompson, Josiah. *Last Second in Dallas*. Lawrence: University Press of Kansas, 2021.

——. *Six Seconds in Dallas*. New York: Bernard Geis, 1967.

Tompkins, Peter. *Secrets of the Great Pyramid*. New York: Harper & Row, 1978.
Trower, W. Peter. *Luis W. Alvarez, 1911–1988*. Washington, DC: National Academy of Sciences, 2009.
Vallee, Jacques. *Forbidden Science: Journals 1957–1969*. Berkeley, CA: North Atlantic Books, 1992.
Waltz, George H., Jr. *What Makes a Scientist?* Garden City, NY: Doubleday, 1959.
Wilford, John Noble. *The Riddle of the Dinosaur.* New York: Alfred A. Knopf, 1985.
Zapruder, Alexandra. *Twenty-Six Seconds: A Personal History of the Zapruder Film*. New York: Twelve, 2016.

INDEX